D0290374

SCOUTING AND SCORING

Scouting
and Scoring

How We Know
What We Know
about Baseball

Christopher J. Phillips

PRINCETON UNIVERSITY PRESS
PRINCETON AND OXFORD

Copyright © 2019 by Princeton University Press

Published by Princeton University Press
41 William Street, Princeton, New Jersey 08540
6 Oxford Street, Woodstock, Oxfordshire OX20 1TR

press.princeton.edu

All Rights Reserved

Library of Congress Control Number: 2018938274
ISBN 978-0-691-18021-2

British Library Cataloging-in-Publication Data is available

Editorial: Eric Crahan, Al Bertrand, and Kristin Zodrow
Production Editorial: Mark Bellis
Jacket Design: Lorraine Doneker
Jacket Credit: © Shutterstock
Production: Jacqueline Poirier
Publicity: Tayler Lord and Kathryn Stevens
Copyeditor: Sarah Vogelsong

This book has been composed in Adobe Text Pro and Gotham

Printed on acid-free paper. ∞

Printed in the United States of America

10 9 8 7 6 5 4 3 2 1

In memory of James William Phillips (1932–1998)

In time, the game will be brought down almost to a mathematical calculation of results from given causes; but, at present, it is merely in its experimental life, as it were.

—HENRY CHADWICK, *THE GAME OF BASE BALL: HOW TO LEARN IT, HOW TO PLAY IT, AND HOW TO TEACH IT* (1868)

CONTENTS

Introduction

In the fall of 2012, Craig Biggio was up for election to baseball's Hall of Fame. A longtime Houston Astro and seven-time Major League All-Star, Biggio seemed like a good candidate. Still, he fell short of the votes required for induction that year, and then again the following year. Supporters of his case put forward evidence to convince the skeptics: numbers concerning doubles hit, batting averages recorded, runs scored; accounts of his loyalty and his apparently steroid-free record; descriptions of his scrappiness and versatility. The debates were not all that different from countless others in which quality is assessed. How do we know how to separate the good surgeons from the bad, or the great teachers from the merely good?

We're told there's been a modern revolution in how we should approach these questions. All humans, we've learned, suffer from unconscious flaws in how we see and think. As a result, we need to gather lots of data about a situation—ideally, numerical data— aggregate them, and analyze them statistically. Only by shackling ourselves to objective data and thereby limiting our own subjective biases and idiosyncrasies can we arrive at reliable knowledge.[1] For Biggio's candidacy, that meant considering his playing statistics, dispassionately assessing his numbers, and comparing them to those of his peers. It wasn't enough to recall him in action or to have fond associations with him: we needed numerical evidence to know if he was worthy of the honor.

The idea that there are seemingly irreconcilable approaches to judging quality in baseball was reinforced by Michael Lewis's 2003 book *Moneyball: The Art of Winning an Unfair Game* and the 2011 feature film based on it. The book and film stirred interest far beyond baseball fans, because Lewis was supposedly describing a general solution to the problem of valuation, especially under financial constraints. Indeed, many took the book as providing a larger lesson. The Harvard Business School used it as a case study in the cultivation of leadership and innovation. Others saw "moneyball" methods as offering new ways to replace tradition-bound fields such as politics (*Moneyball for Government*), law ("Moneyball Sentencing"), education ("Is 'Moneyball' the Next Big Thing in Education?"), and criminal justice ("Lessons for Policing from *Moneyball*").[2]

There's a scene about a half-hour into the film that dramatizes the stakes. Billy Beane, the protagonist and general manager of the Oakland Athletics, enters a room full of scouts trying to fig-ure out which prospects to draft. Beane brings with him his new assistant—an Ivy League graduate with a degree in economics. Beane wants to figure out how the club is going to replace three top players from the previous season who had signed more lucra-tive contracts elsewhere.

The surface issue is money: if the club had enough of it there wouldn't be a problem. But the ultimate question is how to maxi-mize what the team can do with limited resources. The scouts sug-gest they should get the best players they can afford. Beane's new approach—what Lewis called "moneyball"—is to buy outcomes, not players, because outcomes are cheaper. Beane is looking only for runs on offense and outs on defense. In order to rationally al-locate limited resources, Beane argues, they must first turn pros-pects into statistical aggregates. Only the numbers matter.

It isn't difficult to tell who's on which side: The scouts are old; Beane and his new assistant are young. Scouts don't know the word "aggregate"; Beane says they should all be "card counters" at the blackjack table. Scouts forget to carry the one when they calculate; Beane's guy can manipulate numbers on the fly. Scouts know what other clubs think about prospects; Beane knows who gets on base. Scouts want to talk about who is "on the weed" or going to strip clubs; Beane says "on-base percentage" is all they're allowed to discuss. Scouts talk about people; Beane talks about statistics.

This rhetorical division emphasizes a clear distinction between forms of expertise. The scouts think Beane is ignoring the wisdom and experience they represent. They talk about the age and condition of bodies as well as the way players behave on and off the field because ultimately the game is played by fallible humans. Beane redirects the conversation to statistical measures of performance. He wants to know the numbers. The distinction is between scorers and scouts, those who analyze the numbers and those who assess the bodies, but also between analytics and intuition, objectivity and instinct, rationality and superstition.[3] However expressed, scorers and scouts are understood to approach the evaluation of prospects in fundamentally different ways.

Curiously absent from the scene is the fact that every prospect was previously measured and quantified by the scouts themselves. Scouts may not talk about their work as a process of putting numbers on players, but fundamentally that's what they do. Each prospect would have had at least one, and likely many, scouting reports written about him, reports that included scouts' numerical judgments about his present and future abilities. If scouts were so focused on body types and emotional temperaments, it is odd that they bothered to write up reports that calculated a single number for each prospect: an overall future potential, or OFP. At the same time, Beane's numbers come from human "scorers"—from the efforts of fallible statisticians, database creators, and official scorers. They also could not have existed, let alone have been interpreted, without an immense amount of human labor and expertise. The numbers scouts deploy are obviously different sorts of numbers than Beane's, but it still seems strange that there could be such a stark divide in how scouts and scorers approach the world if they both produce numbers that can be used to create a draft list and ultimately put a single price on signing each player. Are scouts' methods really all that distinct from scorers'? The answer, I suspected, would cut to the heart of what it means to produce knowledge useful for making value judgments and predictions in settings far beyond the baseball diamond.

―――――

I am a baseball fan, but I am also a historian. When I began this book I planned only to compare how scorers and scouts evaluate prospects, but I soon realized this was a special case of more general concern in academic fields that occupy my attention. Scouts and scorers document, categorize, and describe the past. They collect data and make judgments about that data in order to make decisions in the present and predictions about the future. Though scorers' and scouts' work is highly consequential, what they do is not all that different from what many of us do everyday: they try to make reliable decisions on the basis of what they know. Though I thought writing about scorers and scouts would be an occasion to release my inner baseball fandom, it turned out instead to be an opportunity for analyzing how reliable knowledge is made.

My topic is baseball, but this is a book about data in the modern world. As the scene in *Moneyball* suggests, not all data are created equal. The numerical data Beane brings to bear on the selection of players are presumed to be precise and objective, and thereby distinct from people knowledge, craft knowledge, or subjective knowledge. Scouting data, conversely, are portrayed as inescapably bound by tradition, culture, and history—that is, bound by the fallibility of humans.

It doesn't make much sense to distinguish numerical data from human data, however, if we think about the word itself. "Data" comes from a form of the Latin verb *dare*, to give. Data are "that which have been given." They didn't originally need to be numerical, objective, or even true. They were simply the principles or assumptions that were conventionally agreed upon so that an argument could take place. Data were that which could be taken for granted. Over time, of course, the meaning has shifted, so that now we tend to think of data as the *result* of an investigation rather than its premise or foundation. In either sense of the term, data take effort to establish and have to be made useful. There is no natural category of "raw" data; data only exist in context.[4]

For Beane and others interested in performance statistics or data analytics within baseball, the primary complaint wasn't that there were no numbers before they got involved, but that the *wrong* sort of numbers had been collected. Since at least the seminal publication of *The Hidden Game of Baseball* in 1984, it has been commonplace to distinguish useful and powerful "new" statistics from "old" or "traditional" statistics.[5] These analysts are quick to remind us that not all statistics are useful, but we often forget the corollary of that assertion: the very act of calling something "data" is a claim about its relevance for a particular argument. "Runs batted in" is a statistic, whether regarded as "old" or "new." It is data, though, only if someone wants to win an argument with it. It's possible, though perhaps mistaken, to imagine facts or numbers existing without people, but it is impossible to imagine data without people.

Books on the new data sciences characteristically spend little or no time discussing the human labor by which data are made. There is often acknowledgement that it matters who collects the data and how they collect them, but the belief, explicitly stated or not, is that with enough sophistication in processing and analyzing, any faults or improprieties in collection can be managed. We can transcend the problems and individual idiosyncrasies of data's origins by collecting enough data.[6] Though it is possible to measure players' abilities or performance without thinking about the origins of the data, if we want to know how scorers and scouts come to know what they know—not just find out who is right— then we have to think more carefully about how they create data.

We can begin by simply asking how facts like Biggio's skill level or hit totals become stable, credible, and reliable. We rarely consider the trust we put in established statistics or how teams come to agreement on whether one player or another should be drafted first. As in any field, technical specifications and practices, politics, education, and social norms shape the creation of knowledge. Yet these factors have become invisible over time. Historian Paul Edwards, in his study of the history of climate models, notes that the difference between "settled knowledge" and a "controversial claim" is ultimately a difference in whether or not the support structure behind each fact is visible. To be a fact means to be supported by an infrastructure, but established, settled facts have made the infrastructure invisible enough that they can seem natural and eternal. Facts are controversial when we can see the infrastructure supporting them.[7] To understand how scoring and scouting knowledge works, I realized I needed to uncover the structures—the labor, technologies, and practices—behind them.

What I discovered was that historically the ways scorers and scouts produced knowledge and established facts were not all that different. Human expertise was required to collect, standardize, and verify performance statistics. Moral considerations determined what data to keep, while complex bureaucratic measures managed scorers' judgments. Scouts were fixated on accurate measures of performance and value. Over time they increasingly had to express their judgments with numbers. They, too, relied on complex bureaucracies and technologies to collect, standardize, and verify their data. Over the last half of the twentieth century scouts and scorers increasingly shared a goal of turning players into numbers.

Any claimed division between scouting's judgment-based subjectivities and scoring's data-based objectivities doesn't have a strong purchase, historically. That is not to say that such distinctions can't be made; in fact, assertions that one process is more objective than another or that one practice minimizes subjective bias can still play important roles in debates. *Moneyball* and similar narratives have presented scouts and scorers as fundamentally divided in part because it makes a good story, a modern-day parable about the power of data and rationality to overcome superstition and guesswork. Parables, like myths, are important cultural markers, as anthropologists have told us for as long as there have been anthropologists. They are ways of organizing social norms and of communicating and maintaining them. But they are not necessarily accurate portrayals of how things work.

Stark divisions between subjective and objective modes, between intuition and measurement, and between different forms of expertise seem inappropriate when we look at how scouts and scorers have acquired knowledge over the years. There are many different ways claims can be made objective, and all of them—trained judgment, regulation and rule following, disinterestedness, mechanization, intersubjectivity, consensus formation—have been used by scouts at one point or another. Similarly, classical markers of subjectivity—judgments of taste and morality, deferral to authority or charisma, management of bodies—have also been applied to scoring practices.[8] Like scorers, scouts are overwhelmingly white and male, and yet they typically treat their bodies as irrelevant to the data they produce, even as their knowledge remains inextricably a product of their own observations. Neither scouts nor scorers care much about such philosophical distinctions, but in practice both groups are deeply concerned with solving the problems of reliably measuring and evaluating people.[9]

One reason baseball is such a good topic for thinking about the practices of evaluation and the nature of data is that performance statistics have been recorded on paper for nearly as long as games have been played, and interested observers have used these records from the beginning to measure and predict excellence. Early clubs—amateur social organizations in which baseball first flourished—nearly always had a scorekeeper and a scorebook in which statistics were recorded. When the sports reporter Henry Chadwick wrote some of the first manuals on baseball in the 1860s, he also noted the importance of scoring, placing the scorekeeper as one of the few people allowed on the field with the players. His desire was that the game of baseball be made American, scientific, and manly, and he believed that the best way to achieve this goal would be through careful recordkeeping.

Just as important as its historic connection with recordkeeping is the fact that baseball analytics has become nearly synonymous with data analytics generally. Nate Silver's rapid rise from independent baseball analyst to the *New York Times* and ESPN's payroll, as well as to *Time Magazine*'s list of the "100 Most Influential People," seemingly proved that thinking about baseball data provides the skills to think about data in many domains. More striking, perhaps, is that baseball is portrayed as an "ideal home" for data analysis. Though otherwise critical of the use of data-driven mathematical models, Cathy O'Neil's *Weapons of Math Destruction* praises baseball's use of statistical algorithms and numerical analyses as healthy, fair, transparent, and rigorous.[10] If baseball is the paradigmatic example of the expertise and benefits that modern data science can provide, then we ought to consider the extent to which baseball really does represent the replacement of one way of ascertaining quality with another.

———

This book is a history of how scorers and scouts know what they know about baseball. The first four chapters cover the history of official scoring and the creation of baseball statistics while the remainder explores the history of scouting. Both parts trace the people, practices, and technologies used to translate the movement of bodies into reliable knowledge. The technologies involved certainly include the high-speed electronic computer, but I am also interested in the more mundane yet pervasive technologies—pencils, papers, scouting reports, stopwatches, and scoresheets—that have enabled data to be collected. Tools like scoresheets and scouting reports are not simply data-recording devices; they *create data* by enabling the relevant aspects of baseball to be made visible and durable.[11] These basic tools are easily forgotten but essential to examining how scorers and scouts know what they know.

This book is neither a thorough history of data analysis in baseball nor a comprehensive account of official scoring or scouting. It is not meant to explore whether stats or scouts are more important to running a baseball team; that it takes both forms of knowledge is obvious to those who manage clubs.[12] Rather, I draw on the history of scoring and scouting, of statistical databases and scouting reports, to show that the attempt to create reliable data about the value of individual players looks quite similar on either side of the claimed scouting–scoring divide. In some ways scouts and scorers make an odd pairing; there certainly are fundamental differences in what they do. Scouts are single-mindedly focused on the future, on finding metrics and heuristics that will enable them to make predictions about who will succeed in the coming months and years. Scorers are more retrospective, collecting data on an ongoing basis while also finding statistics from the past that will help them analyze quality and strategy in the present. Both scorers and scouts, however, are focused on making characterizations and judgments about quality, on finding ways of measuring the abilities of players. Their practices are in many regards remarkably similar.

The book's first four chapters reveal the labor that goes into creating the data behind modern analytical claims. These numbers are powerful, but they are deeply tied to the processes of their creation, collection, and dissemination. These processes have been forgotten or actively ignored in accounts that simply treat the numerical data as reliable and stable. I use "scorers" and "scoring" as broad terms, covering those who are involved in the creation and maintenance of statistical data about baseball, regardless of the end uses to which individuals might apply that data.

The final three chapters focus on scouts. Scouts can perform many different roles for a club, though in general they fall into the categories of professional, advance, and amateur. Professional scouts are typically responsible for evaluating players in the minor leagues or on other teams for acquisition. Advance scouts determine tendencies of future opponents, ferreting out their strengths and weaknesses so that the best strategy can be deployed against them. The focus here will be on amateur scouts, those evaluating nonprofessional players who might currently be in high school, in college, or playing other sports entirely.[13] This is the "hard case," the scouting practice seemingly the furthest removed from scoring; amateur competition is typically so inferior to that of the minor and major leagues as to render the performance statistics of amateurs useless for most clubs.

Instead of emphasizing their differences from scorers, however, I will show how amateur scouts also have tried to make reliable determinations of the value of players. Scouts like to talk about themselves as loners, as renegade hunters looking for diamonds in the rough. But ultimately they are cogs in a giant bureaucratic machine, producing written reports of what they've seen, reports that are turned into quantified evaluations of players. Scouts are hunters of data, recorders of data, and compilers of data. They have elaborate systems of how to see, measure, and evaluate players. They deploy tools and technologies to help quantify skills and ultimately reduce predictions of future performance to a single number.

These chapters rely not only upon memoirs, archival records, and interviews with scouts but also upon the thousands of scouting reports deposited in the library of the National Baseball Hall of Fame in Cooperstown, New York, and placed online in 2014 as part of the Diamond Mines exhibit.[14] This collection is by no means complete, with entire teams and scouts missing from its rolls. Scouting reports are ephemeral—they simply disappear when a general manager or scout goes through his papers and figures that decade-old reports on now-retired minor league players have no value. Nevertheless, the collection in Cooperstown is extensive enough to provide a real sense of how scouting reports have been used over the years. It is just one slice through the history of scouting, but it is a revealing one.

Biggio was elected to the Hall of Fame in 2015, his third time on the ballot. In the end, most commentators considered the decision uncontroversial. The statistics, after all, seemed to speak for themselves: he had 3,060 hits and the most doubles in history by a right-handed batter, just ahead of the turn-of-the-century star Nap Lajoie. After his election, though, I wondered how exactly we knew that Biggio—let alone Lajoie—had precisely that many hits and why we had such high confidence in these numbers. At the same time, I wondered how scouts had seen Biggio, how they had described him and his abilities, and whether they had predicted he would become a Hall of Famer.

No less a statistical authority than Bill James, the author of the influential *Baseball Abstracts*, once called Craig Biggio his favorite player. He explained this opinion in 2008:

> [Biggio] was the player who wasn't a star, but who was just as valuable as the superstars because of his exceptional command of a collection of little skills—getting on base, and avoiding the double play, and stealing a base here and there, and playing defense. Here was the guy who scored 120 runs every year because he hit 45 or 50 doubles every year and walked 70 to 90 times a year and led the majors in being hit with the pitch and hardly ever grounded into a double play and somehow stole 25 to 50 bases every year although he really had very average speed.

James also praised the parts of Biggio's career that didn't show up in the box score, the way his move from catcher to second baseman "required something that you don't often see, an exceptional level of determination, dedication and adaptability." Given the choice between drafting a future Tom Glavine, Ken Griffey, or Frank Thomas, James declared that he would still take Biggio: "Maybe that's not what the numbers say is the right answer, but Biggio was the guy who would do whatever needed to be done.

Makes it a lot easier to build a team." James concluded with a note of sadness that Biggio's career had been like a movie that went on too long. He didn't "admire" the fact that Biggio "hung around" just to get 3,000 hits—it's "like the director can't find the ending so it goes on for another half hour."[15]

Seamlessly—and characteristically—in this summation James wove together numbers and narratives, subjective judgments and objective facts. The original sabermetrician—a term he coined to unite both the "Society for American Baseball Research," known as "SABR," and "measurement"—he refused to make easy distinctions between quantitative and qualitative data. He treated them interchangeably, as reliable, established ways of evaluating a player, and as a basis for making a case for Biggio's worth. If a guru of statistical analysis, one who was supposedly a crucial inspiration for the claim that data analytics should replace traditional ways of judging value in baseball, didn't make stark divisions between scoring and scouting, surely it is worth thinking far more carefully about how both scorers and scouts come to know what they know about baseball.

1

The Bases of Data

"He has the most doubles of any right-handed batter in history." It was a claim confidently bandied about as Craig Biggio was considered for the National Baseball Hall of Fame. And it seemed an objectively true and relatively simple sort of fact—the number of doubles he had hit was greater than the number of doubles any other right-handed player had hit. It was easy to check by heading to baseball-reference.com or some other encyclopedia.

Baseball Reference's list of most doubles hit in a career includes Tris Speaker, Pete Rose, Stan Musial, and Ty Cobb—all either switch- or left-handed-hitters—and then Craig Biggio, with 668 career doubles. But how do we know that Biggio really hit 11 more doubles than the next right-handed hitter on the list, Nap Lajoie? Lajoie played from 1896 to 1916, before sabermetrics, fantasy leagues, and highlight reels—even before radio broadcasts or daily statistical updates. More troubling, at the time of Biggio's candidacy, the powers that be in Major League Baseball disagreed with Baseball Reference and other organizations about whether Lajoie or Cobb had the highest batting average in 1910.[1] Given such distance and uncertainty, how can we be confident Lajoie didn't have more doubles lurking in the records—or that he actually did hit exactly 30 doubles in, say, 1907?

Even a simple claim about performance statistics implies a reliable record of hits going back nearly 150 years. It may seem an objective fact, right or wrong, but that's not to suggest it is a simple or easy thing to be confident about. Believing that Biggio set a record for doubles requires believing in an entire history of recordkeeping and error checking, in an entire structure of people and tools meant to ensure the accuracy and reliability of facts. If we want to figure out *how* we know that Lajoie hit 30 doubles in 1907, then we might as well start by asking where Baseball Reference actually got that number.

———

Baseball Reference's clean interface makes the facts displayed there seem natural, eternal, and indisputable. Biggio's page reveals a dizzying array of numbers, sorted neatly by year and category. Some of the stats provided, like "batting average" and "runs scored," are essentially as old as professional baseball; others, such as "wins above replacement" and "adjusted batting runs," are more recent creations. The interface provides a clever bubble that appears when the cursor is hovered over a statistic, explaining how the number has been calculated. The site even allows users to sum across seasons or other subcategories. The whole structure is geared toward providing a clear display of mathematical certainty. Or, as the founder of the site, Sean Forman, explained, the site's purpose is to "answer questions as quickly, easily, and accurately as possible."[2]

When Forman first put his website online in mid-2000, its ability to generate quick answers was its selling point. Even in this relatively early stage of the internet, there were already other places where fans could find similar data online, including stats.com and totalbaseball.com. These competitors often also had big names, or at least names with authority—totalbaseball.com had signed agreements with *Sports Illustrated*, and stats.com was licensed by a variety of national publications.

The advantage baseball-reference.com offered was a superior interface, which Forman called putting a "friendly face" on existing data. He minimized images and ads, with 95 percent of the pages under 20 kilobytes (kb)—no minor thing, given residential download speeds generally maxed out at 56 kb per second at the turn of the century. Forman had started Baseball Reference as he was finishing his doctoral dissertation on computational protein folding, a field seemingly irrelevant to baseball until he explains that his research was "basically optimization."[3] Forman was good at taking a complicated mess of facts and interconnections, analyzing them, and cleaning them up.

The casual fan might assume Baseball Reference's numbers were coming directly from Major League Baseball or from its official statistician, the Elias Sports Bureau. At the time of Biggio's candidacy, however, Forman had no formal relationship with Elias, and he had never spoken with anyone there. As is the case with many encyclopedias, the specific origins of any given statistic, save a generic note at the bottom of every page, were left unspecified. However elegant its interface, Baseball Reference didn't—and doesn't—provide many overt reasons to trust the statistics that appear there.

As it turns out, Forman had initially taken his data from the statistical database freely provided online in 1996 by another internet-savvy baseball fan, Sean Lahman, at baseball1.com. Lahman, in turn, had built his database using the CD-ROM that came along with the third edition of the groundbreaking encyclopedia *Total Baseball* in 1993. The CD included image files of the entire encyclopedia, with its own reader on the disk to view the individual files. Lahman noticed that the publishers of the CD, Creative Multimedia Corporation, didn't protect its contents very well. With a day job designing databases of digital images for Kodak, Lahman had the skills to post *Total Baseball*'s statistics online for anyone to download.[4]

It's misleading to talk about posting statistics online as if Lahman were simply copying the files from the CD-ROM. A book is just as much a technology for holding and displaying data as a computer file—and perhaps has proved more robust and user-friendly. But Lahman didn't just want to read the book on a computer. Lahman wanted to reverse-engineer a database. He gathered ("scraped") the statistics from the files and then organized them into a relational database by assigning unique IDs to each player, team, and statistical category so that they would be easily searchable. Ultimately, he was able to create his own database, one that relied on the facts as conveyed by *Total Baseball* but that was presented not as the image of a printed table, but as an editable Microsoft Access file.

Lahman decided to put his database online as a result of two frustrations: first, that so many online repositories unexpectedly disappeared in the early days of the internet, and second, that baseball data were often presented in ways that were not conducive to research. Watching Ken Burns's film *Baseball* during the 1994 strike had given Lahman the idea of combining his computing skills with his interest in baseball. In this period before it was common to "surf" the "world wide web," many connoisseurs of baseball statistics had found their way to an active Usenet group: rec.sports.baseball. Lahman found the baseball group a useful resource for sharing statistical research and interesting facts among fans who were also early adopters of computing technology.

However, Lahman soon discovered that it was extremely inefficient to conduct independent research in order to respond to Usenet posts. While it was easy enough for someone to find the answer to basic questions such as the number of doubles Lajoie hit in 1907, it was not easy to find how many doubles had been hit that year by left-handed second basemen or how many doubles had been hit by second basemen over a particular span of years, unless that question had been specifically addressed by an encyclopedia's authors. Someone would have to go through the books player by player, team by team, and year by year. Turning *Total Baseball* into a downloadable database would make Lahman's and other researchers' lives easier: now they could do their own searches and manipulations of the data. The advantage of even rudimentary computerized databases was that they made it possible to query information without needing to collect and reorganize the data every time. A few commands could select all second basemen in 1907, aggregate their doubles totals, and display the results in a matter of seconds. Lahman's database transformed the way baseball research could be done.

In turn, Forman's Baseball Reference was revolutionary because it eliminated any need for skill in database manipulation. By downloading Lahman's database and then constructing a website that would break it up into linked pages, Forman made it possible for people who wanted to know about doubles hit in 1907 to access the data directly through his website. The trick for Forman was writing code such that he could take requests from nonprogrammers, turn them into formal queries on his own server, and then send out the relevant page of data to display on a user's home computer. He could do this by 2000 in a way that Lahman couldn't, or that at least would have been much harder, a few years earlier because an "off the shelf" database management application, MySQL, had not been available until the last half of the 1990s.

Both Lahman and Forman took existing statistics and figured out new ways to display them. If they did not create the data themselves, were they violating copyright by displaying it as their own? It certainly wasn't a far-fetched possibility. Before Lahman posted *Total Baseball*'s statistics online, he had consulted an intellectual property lawyer to get the go-ahead first.

Of course, *Total Baseball* itself had "taken" data from previous publications. There was a tradition extending back more than a century of repackaging and reprinting historical statistics from the Spalding, Beadle, DeWitt, and Reach guides, as well as from the series of titles produced by the *Sporting News*. Who owned the right to print the fact that Lajoie hit 30 doubles during the 1907 season?

As the U.S. Supreme Court stated in 1991, the problem of assigning ownership to data is the "undeniable tension" between "two well-established propositions. The first is that facts are not copyrightable; the other, that compilations of facts generally are."[5] The 1991 case in question was *Feist Publications, Inc. v. Rural Telephone Service*. Both parties were publishers of telephone books—making their dispute a bit dated—and their conflict surrounded the fact that Feist Publications had reprinted over 1,000 listings from Rural's directory in its own book without Rural's assent. The Court found that compilations of facts, like telephone books or statistical tables, are subject to copyright, but not because of the "sweat of the brow," the work that went into the creation of facts. "The distinction," the Court clarified, "is one between creation and discovery: the first person to find and report a particular fact has not created the fact; he or she has merely discovered its existence." Copyright was not a "reward" for the work of compiling facts. Rather, it was an acknowledgement of the creativity and originality of the presentation of facts. In this case, since Rural was not found to have had any originality in its presentation of the facts (it had simply listed entries alphabetically, failing to display the requisite "minimal creative spark"), Feist was not considered to have violated copyright law by reprinting thousands of entries.

On these terms, Lahman and Forman could reprint the data from *Total Baseball* because the facts themselves couldn't be copyrighted. Both Lahman's database and Forman's website took existing facts and packaged them in new and creative ways. The physical presentation of the facts was what mattered. The statistics themselves were treated simply as part of the natural world, separable from any effort that might have gone into their "discovery."

————

Pete Palmer certainly had expended "sweat of the brow" in compiling the statistics that went into the *Total Baseball* CD-ROM in 1993. Palmer had spent much of his youth looking through baseball guides and creating lists from them. He made the kinds of lists curious kids were fascinated by—players with 100 doubles or 100 walks. It was time consuming, but he was that sort of kid. After graduation from Yale in 1958, he used his electrical engineering degree at a series of Cold War–era firms: Raytheon, Sylvania, Mitre, and Systems Development Corporation. RAND, the quintessential postwar research outfit, had spun Systems Development Corporation off in the late 1950s as a stand-alone expert in systems analysis, which by this period entailed a substantial role for electronic computers.[6] When Palmer started at Systems Development Corporation in 1973, his job required him to run a computing mainframe that was linked to a radar installation called Cobra Dane and located at Shemya Air Force Base off the coast of Alaska. Systems Development Corporation had been hired to support the radar computing system, the main purpose of which was to scan the skies for possible Soviet missiles. The computer was needed to calculate whether objects picked up by the radar matched the trajectory of known satellites and planes, or whether they might in fact be enemy missiles.

Palmer's training in systems and electrical engineering not only had landed him this job by the 1970s but had given him new tools to express his disappointment in the state of baseball statistics. He had continued to collect and analyze ever more expansive collections of data produced by *Baseball Magazine, Who's Who in Baseball,* and the *Baseball Cyclopedia.* By the 1960s Palmer had also been able to peruse what are widely regarded as landmark encyclopedias of baseball statistics: *The Official Encyclopedia of Baseball* and *The Baseball Encyclopedia.* The former was published by A. S. Barnes and was less a statistical repository than a biographical index, providing players' names, dates, teams, positions, games, and win/loss records (for pitchers) or batting averages (for nonpitchers). It was, in fact, a collection of glorified roster lists, but—importantly—roster lists that began in 1871. The *Baseball Encyclopedia* was published by Macmillan as the first extensive collection of statistical data, large enough to be known affectionately as "Big Mac."[7]

The creation of the Macmillan guide entailed a herculean effort by a team of researchers headed by David Neft to not only gather information from a wide range of sources but also typeset the final product entirely by computer.[8] In addition to a personal collection put together by Hall of Fame historian Lee Allen, the group drew upon the official records of the leagues. From the start of the National League in 1876, the league office would receive (or, later, designate a statistical agency to receive) a summary of statistics after every game. League rules prescribed exactly which batting, fielding, and pitching statistics were to be recorded. By 1890 the rules, while constantly in flux, specified that the statistical "summary" should contain, in addition to the basic record of hits, assists, put-outs, runs, and errors for each player, the number of earned runs by each side; the number of doubles, triples, home runs, and stolen bases made by each player; the number of double and triple plays and the players involved; the number of walks surrendered by each pitcher; the number of batsmen struck out and hit by pitch; the number of passed balls by the catcher and wild pitches by the pitcher; the time of game; and the name of the umpire.[9]

Once the summaries arrived at the league office, they were entered into large bound ledgers, called "day-by-days" or "dailies," arranged by team and by player so that each line represented a particular performance by a player on a day. The league secretary figured out the season-ending total of doubles for a player by adding up all the doubles hit by a particular player every day throughout the season. In turn, he could figure out the total number of doubles hit by a particular team or the entire league in a season or the number of doubles hit by a player across a career.

One problem for the Macmillan researchers was that the league records only existed for the National League beginning in 1903 and the American League two years after that. So Neft and his colleagues drew from various nineteenth-century "trade papers" for baseball—*Sporting Life*, *Sporting News*, and *Sporting Times*—as well as over 100 local papers from 30 cities that had hosted professional baseball teams. They also turned to a collection of scrapbooks from John Tattersall. In 1941, Tattersall, then in his early 30s and working in Boston's shipbuilding industry, had the foresight to purchase a large number of baseball scrapbooks and sports pages from the *Boston Transcript* as it was going out of business. These dated back to 1876, when a Boston club joined the newly formed National League. Tattersall had first gained attention with his 1953 article in the *Sporting News* on Lajoie's 1901 batting average, having gone through the newspaper files day by day to check the totals given in the official end-of-year records. (His research adjusted Lajoie's average upward from an impressive .405 to an amazing .422.) His methods were notably low-tech—locating newspaper articles, mainly—but, when combined with other sources, they allowed researchers to delve into the earliest records of professional baseball, before the statistical summaries of games and seasons had been standardized.

Despite this impressive midcentury effort to construct baseball encyclopedias, Palmer was questioning their accuracy by the mid-1970s. He had kept his own collection of data, culled from the published guides, and after the Hall of Fame began selling microfilm reels of its dailies, Palmer bought one and used it to fill in statistical details on obscure or overlooked players. (Before he had the microfilm, he would repeatedly have to wake at 4 a.m., drive the four hours from his home in suburban Boston to Cooperstown, New York, sit for eight hours in the Hall of Fame copying the numbers by hand, and then drive home.) Palmer's data seemed to have an increasing number of discrepancies with each new edition of the encyclopedias. Both of the editors of the original *Official Encyclopedia of Baseball*, Hy Turkin and S. C. Thompson, had died, and Palmer kept writing the publisher to inform it that errors were piling up. Likewise, the second edition of the *Baseball Encyclopedia*—and subsequent ones—appeared to Palmer less authoritative and accurate than the first, in part because some players' records had been changed from edition to edition with no notice or justification. Palmer found that these two encyclopedias, despite their vast improvement on prior efforts, didn't match his own data. There was nothing analytically sophisticated about his critique. His lists of numbers simply didn't match theirs.[10]

By 1974, Palmer's pestering had paid off, and he was hired to work on the seventh through tenth editions of the Barnes *Official Encyclopedia*. The first thing he did was computerize its data.

To say that Palmer "computerized" the data makes it sound like a trivial and quick process. It wasn't. In the 1970s, there was no option to optically scan pages, or photograph them and save them as a digital file, or even type them up by hand into a spreadsheet. Computerization required punching physical cards. The technology of the punch card had been around since the nineteenth century and was sometimes called a Hollerith card after Herman Hollerith, who invented them to keep track of data from the 1890 census. Hollerith then made them into a centerpiece of the Tabulating Machines Company (a forerunner of IBM), which he founded a few years later.[11] Computerizing baseball statistics essentially meant turning them into punched cards.

The technology is so foreign to anyone who has worked with computers only after 1985 or so that it is hard to imagine a world in which cards, rather than computers, stored data and programs. In this world, computers physically computed. Long before the term referred to machinery plugged into the wall, it referred to the actual people who did the sums. Computers were originally humans, and usually women.[12] Computing in the early 1970s required a now-unfathomable amount of human labor. In Palmer's case, that meant turning the entire history of baseball statistics— every player's biographical, fielding, and batting history for over a century—into punched cards. By this point, the technology had been widely standardized: 80 columns and 12 rows yielded 960 possible entries on each card. He had to figure out how to turn those 80 columns into the data he wanted to keep—player names, yearly statistical records, team affiliations, and so on. Using a typewriter-style keyboard, Palmer would take a string of letters, numbers, and spaces and translate entries into either a space or a punched hole, which in turn the computer would be able to read and process. (A hole enabled a circuit to be completed and was coded as 1; otherwise, it was coded as 0, turning the hole–no hole distinction into binary code.) A yearly batting line was thus turned into a series of holes and spaces. A single mis-hit key meant junking the card and starting over. Palmer was fairly good at the work, though, recalling decades later that he could punch over 100 cards an hour when he was "in the groove."[13]

For every entry, Palmer recorded the player's name and team, as well as the year; to the basic data of games played and batting average or win/loss record that Barnes provided, he added the league's year-end numbers from the Spalding and Reach guides. By the time he finished, his database took the physical form of a stack of 30,000 or so cards.

Palmer had to design his system so that data could be easily located. He coded an entry in column 43 to indicate whether the card had biographical information (including birth and death dates, as well as a player's full name) or not. He also had to be careful about names, accommodating situations in which two players shared the same name, one player was on two different teams in one year, or the same city had multiple clubs.

The ability to reliably retrieve information from memory across many different datasets is precisely what distinguishes a database from a table of data. "Database" as a term emerged in the mid-1950s, and one of its first uses was in a technical memorandum by Systems Development Corporation, Palmer's employer. "Database" then referred to the physical file system that enabled programmers to interconnect facts into a single set of information. The data *were* the set of cards, and the computer just a complicated device for "processing," or interconnecting, them.[14]

The fact that Lajoie hit 30 doubles in 1907 now existed in multiple forms, from published books to punched cards. It was also ephemeral. There was no place to see or point to 30 doubles across a season outside of these objects. Should two sources not agree, it was sometimes possible to consult other sources, though they might not all be trustworthy. Indeed, Palmer had to sort out the fact that both the Reach and Spalding guides listed Lajoie as having 32 doubles that season, whereas *Baseball Encyclopedia* said he hit 30. Solving the problem of discrepancies, however, didn't require technical skills because it wasn't a technical problem. It was a problem of historical evidence.

The physicality of the fact could not be ignored. The dailies had ink bleed through pages and 1s that looked like 7s. Game summaries were misplaced or lost in the mail, microfilm degraded, and punched cards were corrupted by hanging chads. Printed guides had pages torn out if they hadn't yet been thrown away. Indeed, the New York Public Library has many of the original materials of the history of baseball in its Spalding Collection, but there are huge gaps in even that library's collection of official guides. Many years are missing entirely, and those that exist sometimes have the year-end statistical section torn right out. Websites too were notoriously fickle, especially in the early years of the internet, and commercial websites often survived only as long as the companies that owned them. The site totalbaseball.com simply disappeared from the internet in 2001 as parts of its parent company, Total Sports, were sold. Data are ultimately inextricable from their material instantiations, and yet no storage medium is permanent. The stability of baseball statistics is tied to the durability of physical objects.

Palmer himself was more concerned about the accuracy of his data than their corruptibility, however. He computerized his database with punched cards because he suspected the printed statistics were rife with errors. Even as a kid he had realized that summing the total runs scored by each player on a team didn't always result in the same number a guide reported for total runs scored by the team. The computer enabled him to process his cards and calculate a career summary by tracking players' performance over the years or a team summary by adding together the statistics for each player. Palmer's work on the Cobra Dane radar for Systems Development Corporation and later for Raytheon meant he had access to a CDC 6600, one of the supercomputers manufactured by the Control Data Corporation. The 6600 had been known as one of the fastest computers in the world in the mid-1960s, able to execute about 3 million instructions per second.[15] This was plenty fast for checking baseball statistics.

Once Palmer calculated totals across seasons and players, he could go through and see where the printed records didn't align. There was little reason to suspect perfect correspondence: two newspaper box scores published directly after a game often wouldn't match perfectly, and neither of them might match the statistical summary mailed to the league office, which in turn might or might not perfectly match the day-by-day ledger. But what Palmer's computer did was reveal where the problems were. Perhaps a pinch hitter doubled in his only game but the player had not been entered on the year-end roster. In those circumstances he would not have gotten credit for the double, even though his team might have been credited with it. According to the record, the team would appear to have an "extra" double. (Even Palmer admitted that you've "really got to be a nut to get into this level" of detail.[16]) Palmer noticed that while the National League statistics were fairly reliable—at least their sums of batting and fielding statistics for individuals could be reconciled with team totals starting in 1910 (still leaving more than three decades missing)—the American League didn't even appear to try to match player sums to team totals until 1935, and these numbers didn't reach complete agreement until 1973. Not surprisingly, this was when the statistical operation for the league shifted from the Howe News Bureau to the Sports Information Center—which used computers to check the sums.

Computers are indeed wonderful devices for checking patterns, and in particular for doing arithmetic. Palmer was concerned above all with the consistency of his database, whether it "made sense"—that is, whether it fit together without any "facts" that were logically impossible. What he wasn't concerned about was any sense of ultimate truth in the data. That wasn't the goal of his quest. Without completely reliable daily box scores, let alone official play-by-play data, he couldn't possibly hope to have reliable accounts of every circumstance in which Lajoie hit a double. But he could ensure that his database was logically coherent—that the totals you'd expect to match actually matched.

At the same time that he was painstakingly checking his database, Palmer gained two new ways of verifying his data. First, he began volunteering at (and was later hired by) the Sports Information Center. And by "volunteering," he was really complaining about the mistakes in the records. In return, Palmer got access not only to the newest statistics for the American League, but also to historical records for the American League that he had never seen before.

Second, Palmer joined a new organization of dedicated researchers: the Society for American Baseball Research (SABR). In March of 1971, career civil servant and freelance baseball researcher L. Robert Davids sent a letter around to those he knew had been conducting baseball research, people he called "statistorians" for their combination of statistical and historical interest in baseball. In the letter, he proposed the creation of a more formal group for exchanging information and ensuring that research efforts were not being duplicated. By this point, it was obvious that the task of assessing the accuracy of even a basic statistical account of baseball's history was monumental. Some of the first formal histories of baseball, particularly the work of Harold and Dorothy Seymour, Lawrence S. Ritter, and David Voigt, had appeared in the 1950s and 1960s. But these isolated efforts had not drawn on the legions of fans and assorted baseball "nuts"—in Davids's terminology—who had an interest in the history of the game. Davids himself was motivated in part because the *Sporting News* had discontinued "statistical-historical" articles in the mid-1960s, and he felt there was now no real outlet for sharing this kind of research. On the basis of the response to his letter, Davids decided there was a need for such information sharing and that publication of research results should be a centerpiece of any new organization, whether the research involved photos, parks, birthplaces, home runs, or something different.[17]

The first formal meeting of SABR took place just after induction ceremonies at the National Baseball Hall of Fame in Cooperstown on August 10, 1971. Cliff Kachline, then the historian at the Hall of Fame, hosted the gathering of 16, with Davids elected president.[18] From the start the organization sat on the border between official and unofficial, insider and outsider, meeting in Cooperstown, but not under the control of any of the administrative officials, players, or representatives of the sport itself. SABR was a professional organization without a profession. Its members had other day jobs, and their baseball research, aside from the scattered academics who could justify it as part of their work, was done "on the side." Like Palmer's use of the Cobra Dane computer at Raytheon, borrowed time and downtime are essential factors for understanding how much of early baseball research was accomplished—though Palmer still warns people with a chuckle, "Don't tell Raytheon about it!"[19]

Palmer himself joined SABR shortly after its first meeting. Through the group, he gained access to the efforts of many others devoted to recovering the statistical record for posterity. This expanded network was essential for him, because as much as he could go through the data player by player, year by year, to note the various inconsistencies and search for corrections, there were thousands of discrepancies to investigate, marginal players to rediscover, and midseason trades to track down. Other problems were more particular: the National League records, for example, didn't include sacrifice hits in 1912, making it impossible to figure out how many plate appearances each player had. (A sacrifice was not counted as an official time "at-bat.") Palmer worked with a handful of SABR members to go through newspapers for that year and fill in the game accounts. Other members worked on revising numbers of at-bats, wins, or shutouts. Even something as basic as "games started" for pitchers wasn't recorded as an "official" statistic until 1926 in the American League and 1938 in the National League. Short of looking up newspaper accounts of a game, it could be difficult to tell from the statistical summaries who started a game with multiple pitchers. SABR members might work on any of these problems but were never especially organized because volunteers usually worked only on whatever obsessions motivated them. Palmer admitted that the work of checking historical baseball statistics was too big to be done alone—it required "gremlins all over the country."[20] The monumental task would never have been manageable without SABR volunteers.

Altogether, Palmer has acknowledged by name more than three-dozen SABR members who helped with research, and many more contributed in other ways.[21] Some of this work was incidental to Palmer's database itself, such as that of the SABR Biographical Committee, headed by Kachline of the Hall of Fame, which updated information on players who were not well known or whose information had been listed erroneously in the Barnes encyclopedia. And, admittedly, not all of the work of SABR volunteers was useful. Sometimes a researcher would spend hundreds of hours compiling data only to deliver it in a format requiring dozens more hours to integrate into a database. Or a helpful researcher might not properly track sources, making it impossible to resolve discrepancies.

SABR was not, of course, an organization devoted to Palmer's database. But he had the best database by the late 1970s, and the organization grew with it symbiotically. Upon Tattersall's death in 1981, SABR purchased his collection of scrapbooks and records, which was subsequently maintained by volunteer members. SABR continues to be an effective mechanism for encouraging, directing, and sharing research, and many current researchers are active in publishing with the group or posting to the listserv SABR-L, which was started in summer 1995. (Even then, SABR-L quickly had over 500 members.) However decentralized, SABR became a critical source for verifying statistical data.

When *Total Baseball* was originally published in 1989, with Palmer's database at its core, the bulk of the book appeared as a straightforward set of tables, of facts. The data's reliability, however, was inseparable from the human labor behind the database's creation. Legally, facts are just there, uncopyrightable. In this sense the statistics themselves are treated as natural, regardless of how much "sweat of the brow" went into collecting and creating them.[22] Yet their usefulness depends on the labor that goes into verifying them. Copyright law may only attend to how facts are presented, but without the technologies and people behind them, the numbers would never carry any weight at all.

———

The stability and authority of data, especially records of temporally distant feats, do not come from the fact that the data are numerical. There is nothing inherent about the number 30 that enables it to have stability over time and space. To the extent that data are physical things, they experience friction in movement like all other physical objects, and it takes work to overcome this friction in order to collect, mobilize, and stabilize them.[23]

Once made authoritative, Palmer's database simply could be taken as a collection of facts, and indeed, it was understood to be just that by 1989. Palmer continued to update the database as he incorporated new research—and this is why even though Lahman based his website on the statistics from 1993's edition of *Total Baseball*, his and Palmer's databases increasingly diverged over the following years. When Forman created Baseball Reference, an accurate database was important to him, so he eventually paid up to license the database directly from Palmer and ensure his site would have the most up-to-date version available.[24]

The history of statistical databases is about expertise and credibility, not about insiders and outsiders. "Official" status comes and goes, depending largely on who is granted (or pays for) the privilege of having their statistics labeled as official. The records of the official statistician for Major League Baseball, the Elias Sports Bureau, are notoriously closed to ordinary fans and researchers. Statistics posted on Major League Baseball's official website need not match the data on baseball-reference.com, and reasons for changes to MLB's historical statistics need not be given. Still, one should not overstate the case of insider privilege: Palmer himself had direct access to league statistics from his work with the Sports Information Center, and Elias has published its own widely available books of data in the past. Lahman used the statistics of *Total Baseball* to reverse-engineer his own database, only to be hired a few years later by *Total Baseball*. The most extensive collection of minor league statistical data—now featured on Baseball Reference as well—exists primarily because of the efforts of interested fans, particularly Ed Washuta, Ted Turocy, and others. There is no equivalent "official" historical database for the minor leagues, making this outsider effort as official as any. Moreover, a lack of official status hasn't harmed the reputation of Baseball Reference— far from it. Today, it is the go-to source for many in the major leagues, including those who also have access to Elias; Forman proudly pointed out that a number of club representatives use their official MLB email address to subscribe to Baseball Reference, and that the winter meetings during which baseball executives and staff try to fill vacancies are some of the website's busiest times. He bragged that club representatives have even contacted him to report errors in minor league data—not because it would change the official record, but because everyone treats Baseball Reference as authoritative.[25] Even the Baseball Hall of Fame directs fans' statistical inquiries to Baseball Reference rather than to Elias or Major League Baseball's own website.

Palmer's database has credibility by virtue of the quality of its underlying research and maintenance. Palmer made his database as precise and accurate as possible. He constantly revised his facts in light of new, credible findings. He deployed technology to check for logical coherence and consistency. Most significant, perhaps, he relied upon a community of engaged and self-policing experts. Creating a database of baseball statistics was not all that different than lots of other scientific data collection efforts that draw on hundreds or thousands of volunteers. Whether they involve bird banding, baby observation, or the collection of labor statistics, many projects that require massive data collection efforts blur the line between expert and vernacular knowledge. If we understood Palmer's database as only an inert collection of facts, it would efface the work of volunteers and lay participants, as well as his work of coordination, in the database's construction and verification. Looked at in this light, Palmer's project wasn't on the periphery of "real" scientific work—it was itself a classic example of data-based science. He took many thousands of disparate, even contradictory, historical records and stabilized them into scientific facts.[26]

Recordkeeping may seem at first like a different kind of activity than investigating the laws of nature. Ideas of precision, accuracy, and exactness, however, are inevitably based upon contingent interactions between humans and machines. The distinction between theoretical physics and experimental physics, for example, usually depends on the claim that calculation (exact) and measurement (inexact) belong to different categories. We don't say doubles are "measured": we count them. But counting and arithmetical verification are *always* experimental, and technology based too, whether those technologies are pencils and papers, slide rules, or thousands of punched cards fed through a mainframe computer.[27]

Lajoie's 30 doubles in 1907 were collected on scorers' summary sheets after every game. They were recorded on pages of the day-by-day ledgers at the American League offices, pages that were eventually microfilmed. They were printed in Reach and Spalding guides and in the *Baseball Encyclopedia*. They were put into Palmer's cards and, as the cost of computer memory declined in the 1980s, transferred to a series of floppy disks. They were reprinted in *Total Baseball*'s third edition, and an image of them was included in the accompanying CD-ROM, from which they were copied and posted on Lahman's website. They were downloaded by Forman and uploaded as pages on baseball-reference.com. By the time Biggio was a candidate for the Hall of Fame, anyone anywhere in the world could simply look them up with a few keystrokes.

We only think the medium doesn't matter—that the data are the "same" whether they appear in a book, in a digital file, or on a webpage—because of the triumph of those who wanted to efface the human labor of creating and maintaining facts. In the case of baseball statistics, not all sources agreed on the facts at hand—their stability and reliability had to be created. It is a mark of technology's power that we think of data as able to travel indefinitely with respect to space and time. And when the data are originally numerical, the effect is doubly powerful, for both the content and its format promote the illusion of freestanding knowledge, long separated from fallible bodies and degradable physical forms. It's impressive that we know Lajoie hit 30 doubles in 1907, but it is amazing that there isn't *more* debate about such matters of fact given the amount of labor required for them to be credible at all.

2

Henry Chadwick and
Scoring Technology

It takes a lot of work to make the fact that Nap Lajoie hit 30 doubles in 1907 credible and stable, a statistic that can be called up effortlessly on baseball-reference.com. But such work doesn't explain how we originally knew that he hit those doubles—or, indeed, why anyone bothered to keep count of doubles at all.

It has become commonplace to talk about statistics as if they were natural products of playing the game, detritus that remains after the players leave the stadium, waiting to be collected and saved. But facts don't just appear. They must always be made.

Before the seventeenth century it simply didn't make sense to refer to a natural fact. People might have a theory about the world or knowledge of the world, but there was no conception of particular facts existing *outside* of human agency or ownership. Ordinary features of the world didn't need to be called out with a special designation—they were simply known by experience. Certain experiences, however, were made into a new category of "facts" in order to establish that unusual or strange reports—two-headed cats or extraordinary mushrooms—were credible even if unexplainable. Only in the early modern period did natural philosophers begin using the concept of "fact" to distinguish particular observations from theories that might be constructed about them. Over time, a "fact" came to refer to *any* particular out of which evidence and data could be constructed, providing a common "point of departure" even when people disagreed about how to properly interpret evidence. That we might now nonchalantly deploy the fact of the world's population or the United States' gross domestic product is a historical accomplishment. Baseball statistics could exist as supposedly neutral descriptions of the world only after facts themselves had been established.[1]

The facts of baseball—particularly performance statistics— have played an important role in our understanding of the sport. Allen Guttmann argued decades ago that quantification and recordkeeping were two of the defining features of modern sports. Ancient and medieval sports used numbers, but modernity's singular focus on the "numeration of achievement" and the "quantification of the aesthetic" gave numbers an "extraordinary" status. Baseball was a key example of this phenomenon for Guttmann. Those who had previously looked for why baseball fascinated Americans pointed to its underlying ethos of the "quantified pastoral." In fact, he claimed, baseball "was, in a sense, sociologically primed for the high level of quantification which quickly became a hallmark of the game."[2]

Whatever one thinks of his overall thesis, Guttmann was right that statistical facts were gathered from the earliest days of baseball, many years before the sport was professionalized. There is no need, however, to fall back on claims of cultural "priming" or the influence of a "quantified pastoral" zeitgeist to explain this phenomenon. Baseball's quantifiability was an accomplishment of those who chose to keep track of its history, to take the world of baseball and create facts from it. In this sense, the number of doubles Lajoie hit was never a fact waiting to be discovered. Many other possible facts—the number of foul balls he hit or the number of strikes he watched go by—don't exist at all. That is not to say that Lajoie didn't hit foul balls or take called strikes, but no one bothered to keep track of them. Baseball statistics, like all facts, were made in historically specific ways.

Early scoring in baseball entailed much more than simply tracking who tallied more runs. From the mid-nineteenth century on, baseball fans and clubs kept extensive records of games and individual performances. Scorers were present at baseball games long before the sport was fully professionalized and were seen as playing an essential role in its development. Instead of asserting that baseball is a game primed for quantification, it is more accurate to acknowledge that recordkeeping was a decision that entailed substantial time and effort on the part of clubs. It was an effort that established the scorebook as a technology of fact making, enabling a scorer to reduce the salient events of a game to marks on paper and providing a robust system out of which the statistical facts of baseball could be manufactured.

———

By the late twentieth century, Pete Palmer's database was the most important publicly available collection of these matters of fact. When he and co-editor John Thorn published *Total Baseball* in 1989, they dedicated the book to Henry Chadwick, "the great pioneer of baseball history and record keeping."[3] Chadwick was one of the most important promoters and developers of baseball in the nineteenth century. A British immigrant turned Brooklyn sports reporter, Chadwick became interested in baseball in the late 1850s. Besides reporting on the game he wrote handbooks, served on rules committees and official panels, and eventually edited A. G. Spalding's influential baseball guide. Aside from a brief falling out with the heads of organized baseball in the 1870s, he was central to the game from the late 1850s to his death in 1908.[4]

Chadwick came from a family dedicated to reform through public accountability—that is to say, politically active newspapermen—and in particular to progress through science and rationality. This idea of progress was based on the analysis of objective facts derived from statistical data. Chadwick's older half-brother, Edwin Chadwick, was secretary of the Poor Law Commission in England, and one historian has called his 1842 *Report on the Sanitary Condition of the Labouring Population of Great Britain* a "foundational text of modern public health" in part because of its focus on gathering and publishing extensive statistical tables.[5] Following his father, a radically progressive reformer, the elder Chadwick substituted statistical analysis for medical judgment as the central discipline for understanding the problems of public health. The Chadwicks believed in the power of the press to bring to light statistical facts that would lead to social reform.

Edwin and Henry, though separated by an ocean after Henry's emigration, kept in contact. Each continued to be impressed by the other's efforts. Edwin once joked, "While I have been trying to clean London, my brother has been keeping up the family reputation by trying to clean your sports." Henry in turn wrote to his brother encouraging him to publish the sanitary reports in the United States so that the former colony could also initiate reforms. He believed reformers needed to make the statistical case as plain as possible.[6] Both Chadwicks had faith in social improvement through aggregate statistical analyses, which they thought could reveal the problems as well as the progress that had been made toward addressing them. It was impossible to improve baseball or public health, however, unless one first had accurate records.

Throughout his life, Henry Chadwick was obsessed with keeping score, in the most general sense of the term. Chadwick started scoring baseball games as early as 1860, and his first baseball scorebook has billiard scores in the back, accounts that carefully list umpires for each match, mark the results of every turn, and notate important or unusual events. He also wrote instructions for scoring in tennis. "The only data on which a correct estimate of a player's skills, in Lawn Tennis" can be made, Chadwick claimed, "is that which gives the figures of the score of aces by service and returns." Only with this approach could one gather "the data for an analysis of a player's general skill" to make "a correct average of the season's play."[7] Whatever the activity, Chadwick was dedicated to improvement through statistical recordkeeping.

In baseball, the job of recordkeeping fell to the person acting as "scorer," and for Chadwick, these individuals had an essential, not an auxiliary, role. "Every club should have its regularly appointed scorer," he declared, "and he should be one who fully understands every point of the game, and a person, too, of sufficient power of observation to note down correctly the details of every innings [sic] of the game." The 1859 National Convention rules specified that two scorers—one member of each competing club—needed to be appointed to "record" the game. They were protected by rule from player interference or arbitrary removal and were not allowed to bet on the contest. As early as 1865, diagrams of the game indicated where scorers should stand on the playing field, with their physical presence serving as evidence of their importance. Scorers, like umpires and contestants, had to belong to the game's National Association, or an equivalent state branch, and had to be listed by name.[8] The scorer's role as recordkeeper required him to be a person who was named and known to all participants and obligated to stand by the result as credible. Chadwick knew good data took trustworthy human labor.

Chadwick almost certainly based his account of the proper role of a baseball scorer on his knowledge of cricket, which considered scorers essential participants long before baseball was formalized. By the late eighteenth century cricket matches featured not only scorers, but scorers with pens and paper. Samuel Britcher, scorer for the distinguished Marylebone Cricket Club, began keeping accounts of matches starting in the 1790s.[9] Chadwick himself had scored cricket matches prior to his introduction to baseball.

That's not to suggest, however, that baseball scorekeeping was simply a direct translation of scoring in cricket, aside from the obviously similar "out" and "run" counts. Because there were only two innings in most of the cricket games recorded, each two-page spread was divided by innings, rather than sides, and there was far more attention paid to the performance of the bowler—the pitcher's equivalent—in the cricket match. When newspapers later printed "scores" from cricket and baseball games, similar distinctions applied. This can be seen clearly in reports of games between cricketers and baseball players, in which they would engage in a "doubleheader" of both types of ball game. (Very often these matches were intended to settle the inevitable question of whether baseball or cricket was more difficult, with the ability to win at the other's game being the deciding factor.)

Even as Chadwick's system for cricket scoring evolved over the late 1850s, it never resembled what he would eventually adopt as his approach to baseball scoring. For Chadwick it was not only that baseball was a different game, but that it *ought* to be. Just as cricket was characteristic of "aristocratic and monarchical England," so baseball was of "democratic and republican America."[10] If keeping score was a tool of reform, then scoring baseball was part and parcel of encouraging the scientific improvement of the American game.

———

Despite his clear enthusiasm for the practice, Chadwick certainly didn't invent baseball scorekeeping. Scorebooks are as old as organized baseball. We know this in large part because scorebooks and other club records provide some of the only detailed accounts of games in the 1840s and 1850s. There are extant scorebooks from the Knickerbocker Club of New York as early as October 1845, and from the Atlantic Baseball Club of Brooklyn starting in 1855. We know that the Lawrence Scientific School of Cambridge organized a club and started to play games in late 1858 because its members recorded the games in a handwritten scorebook. Scorebooks emerged alongside baseball clubs, not after them.

The Knickerbocker scorebooks from 1845 dutifully recorded the lineups as well as the outs and runs made by each player.

Within two years, they also recorded the defensive positions of players, as well as the position of players when left on base. This limited approach was the rule: scorebooks in the 1840s and 1850s consistently included only lineups and indications of which players scored and which made an out, giving almost no information about the flow of the game. It is nearly impossible to reconstruct the order of play, let alone the features of each individual at-bat. That wasn't the point. Clubs like the Knickerbockers were tracking individual achievements and failures. If it is possible to figure out who began an inning and who made outs, it is not that tricky to reconstruct how many successful—that is, non-out-making—at-bats took place.[11]

The baseball scorebook emerged in the nineteenth century as a record of meritorious club activity rather than a narrative account of "what happened" in the interest of re-creating a game. Scorebooks included accounts of performance, in that scorers notated runs and outs, but they also included information about the game as an official club event, noting location, day, and participants. The Lawrence Scientific School's beautifully presented book included not only the club's members and games but also its constitution and bylaws. In addition to the lineups and captains, the scorers and umpires were prominently listed. The recordkeeping of these officials was not strictly a matter of counting: one umpire at a Knickerbocker game in the late 1850s noted on the scoresheet that he "cannot comment with too much severity" on the conduct of a particular player. The Knickerbockers also experimented with including a column for fines in the scorebook, presumably for misbehavior on the field. The Alert Club Game Book, a copy of which exists from 1860, similarly began with a simple record of outs and runs but also included preprinted columns for fines and remarks.[12] These scorebooks were analogous to the minutes of club meetings, noting credits and debits of club members.

In many cases it is likely that the formal scorebooks of clubs were created after the fact, based on notes or informal records made during the event. Early scorebooks often had both pencil and pen entries, some likely kept while the game was in progress to make a note of events, and others created as formal documents for the club's permanent records. The Agallian Baseball Club of Wesleyan, Connecticut, was typical in its recording of both practice games and league games in the 1860s. The group kept two different scorebooks for these purposes, one an expensive bound volume and the other a cheaply made book. As with the distinction between accountants' waste books and formal ledgers, scores could first be kept hastily in pencil and then later transferred in pen to formal presentation books. The Atlantic Baseball Club, for example, kept records starting in 1855 but recorded them on preprinted pages that left a space for the date as 186__, requiring recorders of games from the 1850s to cross through the "6" to enter the correct date. Many of the pages have identical handwriting, even though there are a number of different scorers listed for the games. This was not scoring as in-game accounting, but scoring as a permanent record made long after the fact.[13]

Even if the purpose of these club books was largely formal, their structure did allow for basic analysis of data as desired. When the Knickerbockers added a "total runs scored each inning" column to their books in the 1850s, they could effectively reconstruct the consequential action in the sense of who made the first, second, and third outs, as well as who scored and who was left on base in a particular inning (in baseball, every at-bat must result in either an out, a run, or a player left on base). Even the most primitive of books had a way of checking run totals to be sure all were accounted for—the sum over the innings and the sum over the players had to be "proved," or checked to match.

Scorekeeping before the 1860s was largely a matter of the internal recordkeeping of clubs. A baseball club, like all social clubs of the era, required that events be recorded, attendance and fines noted, and honors or demerits tracked. The point is perhaps not surprising, but it is important: by the nineteenth century, baseball culture was a culture of paper inscriptions. A baseball club wrote things down. When Chadwick started covering baseball, he saw the potential of exploiting this internal recordkeeping for reporting purposes. He would eventually take club records and make them into public statistics.

By 1855, newspaper editors were actively seeking baseball clubs' "reports of their play" for publication, a demand that only increased after the creation of the National Association of Base Ball Players in 1857. By this point, baseball had a national following. It was certainly not the first, or even the primary, sport covered by the press. Sports and leisure papers followed horse racing, cricket, boxing, chess, sharpshooting contests, and many other events. Early reporters treated baseball like many of these games, focusing their reporting around the "score" or "box score" of the contest.[14] Before the 1850s were over, however, Chadwick suggested that the press might do more than report on games: it might analyze the sport and its players. On December 10, 1859, Chadwick, writing in the *New York Clipper*, announced the first publication of end-of-season statistics, in which the paper gave an analysis of the fielding and batting of the local Excelsior Club. For each player, Chadwick reported the number of outs and runs made (and averages of each per match), the highest and lowest scores in a match, the number of home runs hit, and the number of matches in which no runs were made, as well as total catches, total outs made on bases, total outs made in general, and outs made per game for each player. He also included "additional statistics": the number of innings played by the club, the number of innings in which the opponents didn't score, the highest score made in an inning, the most passed balls in a game, the longest and shortest games, the most frequent positions played by each player, and more.[15]

Despite the apparent exhaustiveness of his statistical analysis, Chadwick concluded with an apology:

As we were not present at all the matches, we are unable to give any information as to how the players themselves were put out. Next season, however, we intend keeping the correct scores of every contest, and then we shall have data for a full analysis. As this is the first analysis of a Base Ball Club we have seen published, it is of course capable of improvement.[16]

Chadwick clearly was not satisfied with haphazard statistical analysis. He wanted meaningful analysis. On January 14, 1860, he wrote to readers of the *Clipper* that the "manner" in which the scores of 1859 were published made it impossible that a "full analysis of the play can be given." Chadwick had been forced to resort to consulting clubs' official scorebooks, though their inaccuracies made the numbers difficult to tally. Furthermore, he determined that even accurate records were not kept with enough detail to permit analysis of the batting and fielding skill of individuals and teams. He therefore provided an example of tables that should be kept for batting, fielding, and inning-by-inning scoring so that scorers would know what needed to be recorded for the upcoming 1860 season. The *Clipper* promised to devote more space to baseball in the next season, including "full and complete reports" of first-class contests. Chadwick thanked the New York–area teams for welcoming him and promised more information would be published to allow readers to acquire "practical knowledge" of "this excellent and manly pastime."[17] As early as 1860, a decade before any professional league existed, Chadwick was suggesting that proper coverage of the game required a full statistical analysis of play.

Chadwick indeed followed through on his promises, keeping his own extensive records starting in 1860 and pushing the *Clipper* to publish ever more detailed accounts of association games. The *Clipper* established a practice of publishing averages and statistics of many important clubs after they finished the season, giving not only the usual statistics of runs, outs, home runs, and strikeouts, but also "additional statistics," or more detailed analyses.[18]

In this period, Chadwick acknowledged that while the "score of a game" was "the simple record of outs and runs, either of the game or of a player," this basic score was "no criterion of a batsman's skill at all." Rather, the "one true criterion of skill at the bat" was the "number of *times* bases are made on clean hits," followed by the number of bases made, the number of times left on base, and finally the score of outs and runs.[19] By 1860, the *Clipper* was publishing tables titled "Fielding" and "How Put Out" in addition to the scores. Fielding tables indicated the defensive players who made the outs and whether those outs were from catching the ball on the bound or the fly or making an out on the bases. In turn, "How Put Out" tables indicated for each batter whether he made an out on a fly ball, bound ball, one of the bases, or otherwise. (For a period, catches after one bounce—on the "bound"—also counted as outs.) By the end of the decade, the *Clipper* had settled on a standard reporting scheme of a four-column box score, often with outs, runs, bases on clean hits, and number of times left on base after clean hits. This provided the major classes of each at-bat: good hits that resulted in runs, good hits that resulted in men "left on base," and outs made. Though the *Clipper* and other papers fluctuated in their use of this system, by the 1870s, the typical score reported at-bats, hits, runs, assists, errors, and put-outs for each game. Everything else was listed after the score as a special event (hit-by-pitch, walks, flies caught, and so on).

Papers like the *Clipper* promoted these analyses in part because they provided fodder for discussions about year-end laurels. Then, as now, lists ranking top players were popular because they created controversies and increased sales. At the same time, such awards encouraged teams to keep accurate records. Credible rankings required good data, Chadwick repeatedly reminded clubs.[20] Such accuracy was in everyone's best interest: failing to keep records in the desired way meant clubs and individuals simply wouldn't be counted and their achievements would be rendered invisible in postseason awards.

The *Clipper* and other papers soon began listing top players or regularly ranking players by batting average and crowning one of them as the batting champion. Though Chadwick noted that sometimes it was "no easy matter deciding on the respective merits," he was nevertheless happy to contribute to the debate. By the end of the 1860s, the *Clipper* not only sponsored "gold ball" and "silver ball" championship matches but also devoted long articles to the previous year's best players. The "Clipper Gold Medals"— nine of which were awarded for 1868, one for each position—were bestowed on the basis of research and "official information."[21] In determining who would receive these medals, the newspaper asked for team records and also consulted its own scorebook. Reliable data had become essential.

The 1872 *De Witt's Baseball Guide*, also edited by Chadwick, represents the culmination of these developments, including extensive coverage of "noteworthy" contests and "model games" as well as records for each club and reviews of the previous professional and amateur seasons. Nevertheless, more than a decade after setting out to provide clear analyses, Chadwick was still complaining in 1873 that the tables were not adequate because "many games were played without a properly-kept score, and consequently in these no account of base hits were recorded."[22] But in general, systems for the collection of data for the analysis of play had been firmly established by the time the first professional league, the National Association of Professional Base Ball Players, was created in 1871.

———

For Chadwick, systematic analysis meant the systematic creation of data. Accurate estimates of skill required that games be "recorded in a uniform manner," and to this end, he often included a copy of his own scoresheet in his publications, starting with *Beadle's Baseball Guide* for 1861. He claimed that this form was "generally adopted" by baseball clubs, although that was likely more salesmanship than a matter of fact. His goal was to promote a uniform scoresheet, one that would enable not only "ordinary scoring" of outs and runs but also the creation of "detailed reports" of how players were put out and by whom. What he needed above all, though, was consistency— hence his desire to standardize and sell his scoresheet. Only then might "full" reports be acquired from all clubs.[23]

THE SCORE OF THE _____ BASE BALL CLUB OF _____

Base Run'g	Bases on Err'rs	Bases on Hits	Totals.	Time Play Called.	When Played 186_ INNINGS.										Time Game Ended.	Good Plays in Field.								
H	L	B M	T	B	O	R	Batsmen.	1	2	3	4	5	6	7	8	9	10	Fielders.	P	B	F	L	D	K	R	T	A
							1											1									
							2											2									
							3											3									
							4											4									
							5											5									
							6											6									
							7											7									
							8											8									
							9											9									
							Totals.											Totals.									
						Grand Totals.																					

WINNING CLUB _____   SCORER _____

UMPIRE _____   DURATION OF GAME _____ HOURS _____ MINUTES _____

FIGURE 2.1: Chadwick's preprinted scoresheets were meant to spread his system to scorers around the country. Chadwick, *Base Ball Manual* (1874), p. 34

Chadwick considered his own method of scoring a vast improvement on previous attempts to keep track of the outs and runs made. As he congratulated himself in 1868, "My first improvement" to the game was "an innovation on the simple method in scoring then in vogue." His own system, he claimed, measured individual and collective achievement by recording how the ball was hit, who fielded it, and what happened to the batter. In 1861, he advertised his clever use of a single number, derived from a player's position in the batting order (i.e., one through nine), to record actions on the field.[24] The batting order was fixed, so "2" would reliably refer to the second person in the team's batting order throughout the game, regardless of where "2" was actually playing (though Chadwick would also typically indicate defensive position in the batting order). Instead of noting that the batter hit a ground ball to the first baseman, who tagged him out, Chadwick indicated that the batter hit to Smith, the second person in the opponent's lineup, who tagged him out at first base.

This approach focused attention on individual accomplishments and failures. Wild throws were distinguished from missed catches (instead of simply calling both an "error"), and wild throws above the head were distinguished from those on the ground. Chadwick even advocated a slightly different symbolic notation to differentiate significant from minor errors. Each batted ball's "character"—whether it was a bounding ball, a ground ball, a fly ball, or possibly just a "poor" hit—was also noted. There is little evidence he ever regularly deployed this system in full—or that anyone else did either, for that matter (even the most elaborate of Chadwick's own scorebooks seem to use only a partial version of his system). However aspirational, it proved a point for him: the goal of scoring was not to record the way a team won or lost a game, but to analyze the character of each play, to assign credit and fault to those involved.[25]

By the end of the 1860s, Chadwick was claiming that the system of abbreviations that enabled a scorer to notate the result of at-bats "includes everything necessary to record the important details of a game."[26] With false modesty, he regretted that his own scorebooks looked like "hieroglyphics." Deciphering his account books, though, was not about reading a foreign language but about making an analysis of the advancement of the game and gaining practical knowledge of it.

Chadwick was quite concerned throughout the 1860s with figuring out how best to reduce games to numerical tables in part because he believed that the reporting of statistics would affect the sport itself. This wasn't an obvious claim. Certainly, one could imagine the *Clipper*'s reports having no effect on league play—that is, failing to translate into aggregate changes in the game. But Chadwick thought otherwise. In Chadwick's terms, through analysis, "not only the estimate of each player's skill can be made, but the progress made in the game itself will be indicated." Newspapers could "promote the welfare of the game by making these

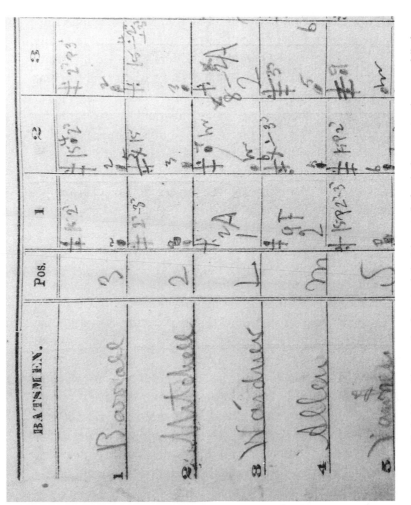

FIGURE 2.2: Chadwick used his scorebooks to track the "character" of each play, warning they were full of "hieroglyphics." Henry Chadwick Papers, Spalding Collection

analyses."[27] Chadwick claimed that the decisions that went into what to track and what to publish would affect the way the game was played and, ultimately, whether it was improved.

One example of his faith in the effect of reporting statistics was his equivocation about whether or not to publish fielding errors. Though he initially kept track of errors in his scorebooks, at one point he recommended eliminating columns for fielding errors, noting that if good plays were enumerated, it was "punishment enough" for players to have "few good marks for skillful play." If errors were published, he argued, it would have a "far more injurious effect," because players would then be more afraid of making poor plays than making a "strenuous" effort in hopes of a good play. He went back and forth on this point over time but clearly ascribed real consequences to the reporting of errors. In his mind, recording and publicizing statistics could change the way the game was played.[28]

Sometimes statistics simply helped put the relative value of plays into perspective. If one wanted to know the frequency of singles as opposed to doubles and triples, one only had to consult the scorebook: the "chances for making more than the first base by clean hits, decrease in proportion to the number of bases the batsman tries to run, the first base being made by clean hits three times to the second's one, and six times as often as the third is." The left fielder needed to be "a good runner, a fine thrower, and an excellent and sure catcher; as probably three out of every six balls are sent toward the left field." In contrast, the right fielder could be "the poorest player of the nine—if there be any such," since "it is only occasionally, in comparison to the other positions of the field, that balls are sent in this direction."[29] Keeping score facilitated strategic decisions.

As baseball's rules changed over the course of the nineteenth century, Chadwick was always concerned with thinking about improvements to them, based upon his own experience and analysis. He marked up his own copy of the rules from December 1868 with possible improvements. Indeed, when trying to specify rules for what should count as a base hit in the 1870s, Chadwick deployed statistical reasoning. When a hit ball rolled slowly into the triangle of pitcher, catcher, and first baseman and the batter reached first base safely, was that the result of an error or a hit? For Chadwick, the debate came down to the fact that in the 30-odd instances he had seen this situation, the batter-runner was put-out in only four cases, and even then mainly because of slow running. Thus, he labeled this event a hit, since one wouldn't want to punish fielders for being unable to make what was apparently a difficult play.[30]

This kind of conclusion required statistical records: simply looking at one case wouldn't have indicated whether it was a routine play or not. In the early days of the game, it was not known whether a successful hit was a result more of the batter's skill or of the fielder's failure. Chadwick put forward the case that simply watching the play would never be sufficient. A scorer also needed to know what happened "in general." That's another reason why statistical records were useful.

Chadwick's criticism was not always successful. As modern baseball analysts have noted with some derision, Chadwick approved of changing the rules to allow 10 fielders and yet disapproved of home runs. As always, though, he was making a claim about the role of data in revealing long-term trends. Home runs look great— they score runs immediately and bring renown to players. By the late 1860s, though, he believed that consistent singles were the best way to show ability and durability over the course of a season, therefore proving oneself more valuable to the team. As he wrote in 1868, "Players who engage in matches frequently have learned the value of economizing their strength. Long hits are showy but they do not pay in the long run. . . . Take the average of heavy hitters, and you will find" that the singles hitters "possess the best record at the close of the season." To give credence to this contention, he cited the example of a player who changed his approach away from hitting long balls and whose record was now "the best ever credited to him." Chadwick claimed the data showed the underlying reality obscured by the "showy" short-term success: the players who hit for singles consistently had better long-term performances. We need not credit this as a true statement (or absolve him of his sermonizing) to understand that he was making a point here about the ability of data to reveal hidden facets of the game. Even in his failed attempts to change the game, Chadwick made claims on the basis of statistical experience, which required careful records of games. Chadwick's claims took a form that was and is a standard ploy of those who keep data: he maintained that the only way to challenge his recommendations was to keep "an equally detailed analysis."[31]

Ultimately Chadwick's analysis focused on how to manage problems of judgment. He was skeptical of measuring batting ability by counting total bases earned on hits, because he felt this measure was "not as reliable" as counting the number of hits since mistakes are easily made in determining how many bases beyond the first are "earned" by the batter. While he was willing to accept the judgment of hit or error for a specific play, he felt it required "far more care and judgment to determine" the number of bases that could be credited to the batter's ability. Clean hits, not total bases, became Chadwick's standard—easy to record reliably, and easy to keep track of consistently.[32]

Chadwick claimed his scoring system had other virtues as well. It emphasized the balance of play: defense must be the opposite of offense. Consequently, he designed his system of scorekeeping on the basis of double-entry bookkeeping. Each play was entered in both defensive and offensive records; at the end of the game, the total events had to balance. One team's hit was the opposition's hit allowed. One team's outs were the other's put-outs. Moreover, the scorebook was clearly set up to facilitate summing across rows (total individual performance) and summing down columns (total team performance in a particular statistic). One could quickly check whether doubles hit by all players summed to the number of doubles hit by the entire team, as well as the number of doubles given up by opposing pitchers, or whether put-outs and offensive outs summed to 27. These were explicit choices meant to emphasize scoring as rational accounting, a precise technical system with self-correcting mechanisms used to legitimate decisions.[33]

This use of accounting also had a moral dimension. Nineteenth-century American accounting practices were replete with attempts to gauge a person's "moral" value and credibility through detailed recordkeeping. Accounting clerks would use their ledgers to collect and view data about individuals' and companies' creditability and thus their reliability for future loans and risks.[34] Chadwick's baseball writing—like his half-brother's work in public health—followed the model of the contemporary Belgian social scientist Adolphe Quetelet in emphasizing that natural and social phenomena, from the movement of the planets to murder rates, exhibited statistical regularity. Unlike the planets, however, the social order could be improved by finding and altering the causes of undesired effects. As Chadwick explained in 1868, in terms that directly echoed Quetelet's, "In time, the game will be brought down almost to a mathematical calculation of results from given causes." His acknowledgment that baseball was now "merely in its experimental life" was an indication that more data needed to be collected before any improvement in results might be made.[35]

Chadwick's vision of progress first required ascertaining statistical regularities and averages for both undesirable and desirable behavior—precisely the work of baseball scorers. The practice of assigning credits and debits was at once a practice of numerical accounting and one of moral judgment. In his instructions for the game, he was clear that fielders' put-outs and assists were mainly about which player *should be credited*. Likewise, when the fielders erred, he maintained that the batsman might make his base safely but was "not entitled to the credit of a base on a hit."[36] He noted that all balls muffed in the field (and a chance for an out missed), even if foul and without a runner advancing on the bases, must be indicated in the scorebook. Similarly, scorers were encouraged to add a special mark to the record when the batsman "hits a fine ball deserving two or more bases" but, out of "good fielding, a fear of

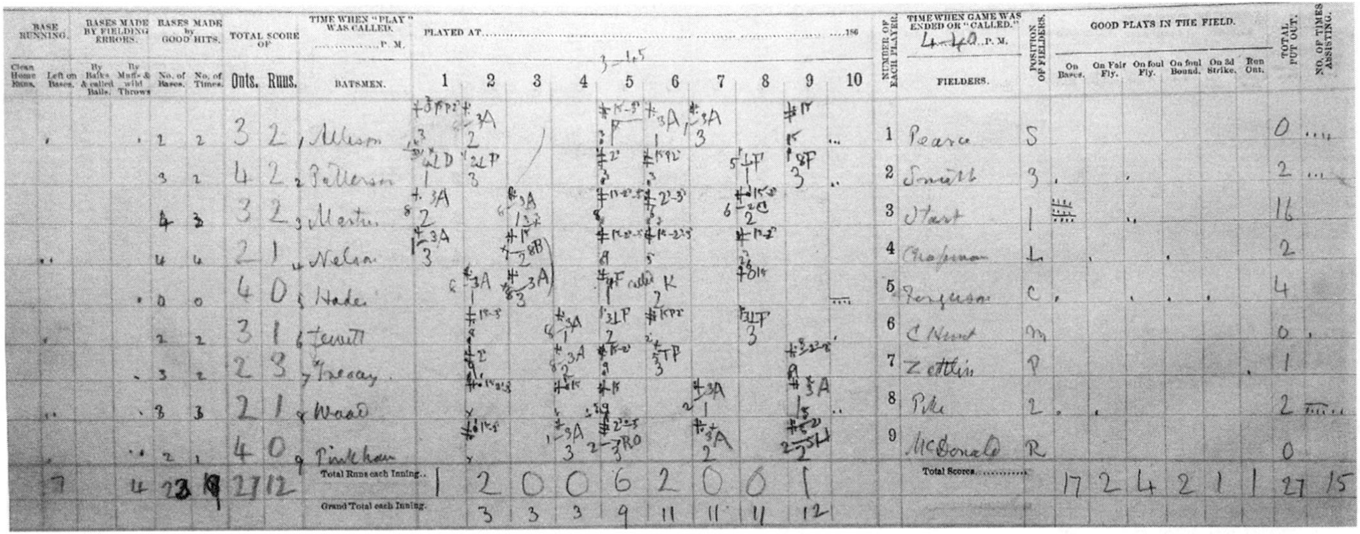

FIGURE 2.3: Chadwick's scorebook not only distinguished "good plays" and "good hits" from "errors" but also mimicked double-entry bookkeeping. Henry Chadwick Papers, Spalding Collection

consequences, or laziness," stops on first base, thus scoring only a "single."[37] The resulting statistical data could serve two purposes simultaneously: it could "stand in" for particular players, enabling a player to be assessed by his total of hits or errors, or it could be aggregated, enabling a scorer to compare two teams' overall performance or trace one team over time to see if improvement had been made.[38]

For Chadwick, proper scoring did not distinguish skillful play from moral accountability: "good" play on the field was always manly, rational, and virtuous. Columns for marking plays in some of his scorebooks divided "good" plays from other plays, so that one tallied a good play in the field (e.g., a put-out, an assist) in one section and a poor play (e.g., an error, a misthrown ball) in another. The language of baseball had long carried moral overtones—most notably in distinguishing "fair" and "foul" balls—but Chadwick sought to make the character of *every* action visible in the scorebook. In this way one would always know the worth of a particular event and thus be able, at a glance, to perceive ability and skill.

These were *choices* Chadwick made. Idiosyncratic at best, and arbitrary at worst, these decisions were meant to create a moral record of the game. He could have chosen to ignore whether or not fielders acted honorably and simply record how many bases were acquired; he could have chosen to ignore the number of bases and instead record solely the distance the ball was hit; he could have recorded the type of pitch (high, fast, low, etc.) rather than its outcome. But he did none of these things.

Another way to see his system as a choice is to recognize how much of what happened on the field he missed. He recorded nothing of players' utterances or adjustments of clothing or of events irrelevant to the play at hand. Many aspects that might actually have been important to players and managers—unsuccessful pickoff attempts, aborted steals on foul balls, changes in defensive positioning—were ignored. The choice not to record an act also had a moral valence: a judgment of worthlessness is still a judgment.

Chadwick promoted a system that, above all, was concerned with making visible and accountable the apportioning of credit and blame. He made the scoresheet into what historian Jules Tygiel called "a series of mini-morality plays."[39] He was not only interested in the labeling of certain acts or players as skillful. He also wanted to enact moral reform through statistical record-keeping. Chadwick did not distinguish scientific progress from moral progress; his scorebook was a scientific record of moral credit.

This emphasis on morality was in line with much of Chadwick's early sermonizing, his attempt to convince skeptics that the game was appropriate for even the most discerning spectator. One of Chadwick's favorite observations in the mid-1860s was that even a "fastidious moralist" would not find fault with baseball as either a "desirable means" of exercise or an "exciting game." Baseball was appropriate for ladies and endorsed by clergy. Similarly, when Chadwick described the "model" baseball player, he drew the portrait of a person who not only was able to physically catch balls and run bases, but also avoided disputes with umpires, followed rules, kept his temper under control, and refrained from profanity and gambling.[40]

For many early observers of the game, Chadwick included, one of the major threats to the sport came from gambling syndicates and others who wagered on contests. The need to control gambling, even as proponents recognized the increased attention and attendance it brought to the sport, was indeed central to early reformers' worries. One historian, John Thorn, has suggested gambling was integral to the quantification of baseball, of the keeping of statistical records of all sorts. Thorn links gambling to the need to have accurate, public records of games to provide gamblers clear measures upon which they could wager.[41] The irony was that two morally opposed groups could both desire quantification of the game and standardization of its data. Chadwick may have had a particular moral purpose in mind, but that certainly did not preclude his statistical project gaining adherents because it enabled an activity that Chadwick himself rejected on moral grounds.

Though Chadwick played a crucial role in the establishment of scoring procedures and the standardization of the scoresheet, there were many others involved as well. There are, in fact, extant records of far more elaborate systems for keeping score in this period than Chadwick's. The famous Red Stockings club maintained a scorebook from the late 1860s—singled out for praise by Chadwick himself, who claimed the Cincinnati club's "efficient scorer" had "one of the most elaborately prepared books we have ever seen"—that recorded how every base was acquired, the character of the play by which all runners advanced, and the quality of umpires' decisions. There were also rival scoring systems, including one of another reporter, M. J. Kelly. Former Red Stockings player Harry Wright's scorebook was innovative in using numbers to refer to positions on the field instead of batting order (so an out made by "1" would be made by the pitcher rather than by the fielder who batted first).[42] Wright's system would, in the long run, become even more popular than Chadwick's.

A number of commercial scorebooks appeared in this period, including *Peck's New Pocket Base Ball Score Book* and *Peck and Co.'s New Association Field Score Book*. The price of the former ranged from 10 cents to 88 cents ($2 to $8 in today's money) and that of the latter from $2 to $15 (at least $30 in today's money), depending on size and format. Some teams had their own book; the Boston Base Ball Club, for example, used Harry Wright's design.

Of course, regardless of the scoring system or book used, scorers and clubs might simply have kept the data they wanted to keep. Just because a column was present or instructions given doesn't mean they were filled in or followed. There's some historical evidence, in fact, that clubs scored as they wished, regardless of format. The Mutual Club game book from 1871 used Chadwick's system, but the scorer often ignored the majority of columns, instead focusing on just runs and outs, as the group presumably had done before adopting Chadwick's forms. In 1881, on the other hand, a scorekeeper crossed out the columns for outs and runs and replaced them with columns for runs, hits, put-outs, and assists, the four statistical categories that were most widely reported at the time. That's not to say there's no significance in the scorebooks' structure; Chadwick and others clearly wanted to have scores kept in a particular way. Rather, it's important to note that preprinted forms can only discipline users so much.[43]

As early as 1869, Chadwick worried about all the variation in scoring styles then in use. One system of scoring should be adopted, he claimed, and he recommended his own. Chadwick's claim against most other systems, tellingly, was that they didn't produce the right data. He noted that "the great majority of scorers possess no regular method of preserving the data of each season's play in such form as to admit of a correct analysis of a season's batting and fielding of each player of a nine."[44] Proper scorekeeping techniques, in his view, were essential to the "correct" analysis of ability.

Others contributed to the establishment of scoring and statistics as an integral part of baseball, but Chadwick deserves the bulk of the credit during the 1860s for ensuring that scorers were included in the playing of every game, that scorers had access to standardized scorebooks and practices for inscribing the action on the field, and that this action was increasingly collected and maintained by associations and leagues. By 1865, a resolution had been adopted by all clubs in the National Association pledging that club averages for the year would be maintained and published. In response, clubs began to calculate averages and circulate them during the season (if they had not been doing so before).[45]

An industry of baseball statistics was springing up around the country. Chadwick happily noted that by the end of the 1860s, "some of the best scorers in the country took pains to prepare model documents in the way of analytical statements of the season's play."[46] It was clear that the coordination of many dedicated scorers was needed to ensure accurate statistics. Only then might baseball become a game of numbers.

Long before the sport was professionalized, baseball was immersed in data. Alongside many other nineteenth-century institutions, from asylums to census bureaus, baseball statistics formed a part of what philosopher Ian Hacking once evocatively called the "avalanche of printed numbers." The appearance of statistical records, Hacking noted, was only the surface effect of new bureaucracies and techniques for classifying and counting. The metaphor of the avalanche regrettably lends a sense of inevitability or natural momentum to these changes, even though people like Henry Chadwick had to be indefatigable in their efforts to promote scorekeeping within baseball.[47] In turn, these practices and technologies of scorekeeping led to the production of written records that would give the appearance of a rule-governed system from which modern facts could be extracted. People could and did disagree with what was counted and how it was counted. But by the 1860s, nearly everyone in baseball was counting in one form or another.

It is not the case that nineteenth-century statistical practices were moral whereas modern decisions are rational. Statistics always emerge from human decisions and moral judgments about what *ought* to be counted. Admittedly, Chadwick's concerns differ widely from those of today's fans. That he downplayed the importance of walks and even railed against home runs was in part a product of the different practices of pitching and hitting that were prevalent in the mid-nineteenth century. But they were also a consequence of Chadwick's own vision of the proper way to play baseball, and in particular of the effect that the moral accounting of individual performance had on the overall development of the game. He wished to make the game more "manly" and "scientific," and to do that he encouraged the awarding of moral credits and debits, the aggregation of them across seasons (and careers), and the analysis of them as data for the measurement of progress.

Objective matters of fact—that 30 doubles were hit by a single player in a single season—are constituted in part by the presence of scorers, the technologies of scorebooks, and the decisions of newspapers, fans, and leagues about what exactly should be recorded. It may be possible to communicate or display these facts as natural aspects of the world, but they were created (and others were not) because of specific historical circumstances. The statistics we have inherited are imbued not just with the human labor of their collection and maintenance, but also with decisions about what should be counted, measured, and valued in the first place.

3

Official Scoring

The technology by which we know Nap Lajoie hit 30 doubles in 1907 is quaint: just pen, paper, and the post office. In the early days of baseball, after each game, a statistical summary form, including a record of any doubles hit that day, was completed and mailed to the league office, where it was written down in a daily ledger of statistical performance across the league. Though technology had changed dramatically by the time Craig Biggio hit 31 doubles exactly a century later, we knew this fact because of basically the same process. An official scorer wrote down, on paper, the number of doubles hit each game and sent the sheet—then by fax—to the league statistician, who summed the totals across the season.

Henry Chadwick and Pete Palmer may have collected baseball data, but the data originally came from an official scorer. Whereas some statistics are manufactured by umpires—called strikes, caught fly balls, home runs—others, like the double or the wild pitch, are essentially created by the scorers. The distinction made is that umpires are in charge of enforcing the rules of the game while scorers are in charge of making judgments about credit. An umpire might determine that a player beat the ball to first base and was therefore safe; a scorer judges whether that action should be

HOME CLUB

BATTERS	Pos	AB	R	H	Total Bases	2B	3B	HR	RBI	Sacrifice		HB	BB		SO	SB	CS	GIDP	PO	A	E	DP
										Bunt	Fly		Tot	Int								
TOTALS																						

PINCH HITTERS (tell how)	Name	Inning	PINCH RUNNERS	Inning
A	for	in	1 ran for	in
B	for	in	2 ran for	in
C	for	in	3 ran for	in
D	for	in	4 ran for	in
E	for	in	5 ran for	in
F	for	in	6 ran for	in

Number out when winning run scored

1st Base on Interference

Passed Balls

Home Run With bases Full

Double Plays

Box Score Proof

Runs _____ AB _____

LOB _____ BB _____

Opponents Put Outs _____ Sacrifices _____

HBP _____

1ˢᵗ on Int _____

Totals must be equal.

Total	Total

PITCHERS

PITCHERS	W L S	IP	H	AB	Batters Facing Pitcher	R	ER	HR	Bunt	Fly	HB	Tot	Int	SO	WP	BK	Start	Finish
TOTALS																		

(Sacrifice: Bunt / Fly; BB: Tot / Int)

Scoring by Inning

Scoring by Inning	1	2	3	4	5	6	7	8	9	R	H	E
Visiting Club												
Home Club												

Official Report of Major League Baseball Game

Played in the city of _____ on _____

_____ League

_____ vs _____
Visiting Club   Home Club

UMPIRES

_____

_____

_____

_____

Official Scorer

_____

Game Time

Started at _____

Ended at _____

Elapsed time _____

Weather _____

Grounds _____

Attendance _____

FIGURE 3.1: The official scorer's form required an enumeration of fielding, pitching, and batting statistics; there was no such thing as an "official" box score or play-by-play account. Courtesy of Major League Baseball

recorded as a single (a credit to the batter) or as an error (a fault of the fielding team). The movements of baseball players around the field are unpredictable yet highly choreographed. Umpires and official scorers determine how these movements are translated into statistics. Only the scorers keep the official written statistical account of the game, however, the account from which all league statistics will be derived. The end result may be an array of numbers on a page, but it is an act of the scorer's judgment that essentially creates the statistics.

Fans have usually paid little attention to this crucial role of official scorers. Indeed, as journalist Bill Plaschke noted in a splenetic 1993 *Los Angeles Times* article, when the interested public discovered that even in the late twentieth century scorers were poorly paid, little trained, and only informally monitored, they were rightfully appalled. How were fans to know that the statistics created were reliable? The only attempt at explicit quality control in the recent past was an open-book test, an exam Plaschke thought was so pathetic that he quoted one scorer's view that "only an idiot couldn't pass it." Plaschke concluded that for the "caretakers of baseball's sacred scoring rules, there are no rules."[1]

Plaschke had a point: it *is* strange that even as baseball statistics have become stable, objective facts, baseball fans have seemed to care so little about the process by which they are made. It is admittedly not obvious who is in charge of official scorers. Starting in the 1980s, they were managed by Phyllis Merhige, who retired in 2016 as the senior vice president for club relations at Major League Baseball (she shared responsibilities for part of that time with Katy Feeney, who died in 2017).[2] During this period, nearly all scorers were nominated by teams' public relations (PR) directors or recommended by current scorers—there was no general call for applications or any official screening process. Aspiring scorers basically had to know the people already in the press box. While Merhige would take a look at applicants' resumes, the key test was in-game experience. Merhige would ask the local PR director to invite a prospective scorer to shadow the official scorer for a few games and give her a report. If the report was favorable, then Merhige sent the scorer a letter of appointment, a rulebook, and the basic guidelines of scoring, and would assign him or her a few games as a test. (There was indeed once an open-book test for screening potential National League scorers, but it was eliminated after the leagues merged in 2000.)

Since all scorers were considered outside consultants on annual contracts, they served at the pleasure of the league—which, as Merhige clarified, actually meant they had to keep the local clubs happy, since she would call the PR director or the appropriate club liaison first if there were complaints about a scorer. Evaluation of a scorer's judgment was therefore partially dependent on whether the local PR director thought he or she was doing a good job. Scorers themselves often walked a fine line, quick to note that they were employed by the league and not the club, but also clear that good standing with clubs was essential. One official scorer, Ben Trittipoe, described the ideal relationship as having the club accept a scorer as "one of their own." In this exchange, the league acted as coordinator—Merhige averred that she was "not an expert on scoring," "just an expert on people." Scorers didn't even submit reports to her at MLB's offices; rather, they submitted them directly to the Elias Sports Bureau, the designated official statistician. Major League Baseball might have been responsible for official scoring but was essentially not in the business of finding scorers, evaluating them, or training their judgments.[3]

Merhige admitted the system wasn't perfect, but she thought it was good enough. Scorers, after all, were not easy to find: they didn't get paid a lot of money per game (about $180 in 2017), they couldn't really take a break or even take their eyes off the game, and they constantly ran the risk of having to navigate the high-pressure situations of hitting streaks, milestones, and potential no-hitters. There were good financial reasons for having a relatively unknown person do the scoring. And while a few were effectively fired by not being asked back the following year (fewer than one a year on average since 2000), the system was surprisingly stable.

Still, even Merhige got frustrated with official scoring's ad hoc nature. For her, the problem was that there seemed to be no effective way to standardize scoring. Individual arrangements for official scorers were anything but consistent—some cities had one scorer doing all or most of the games, while others had a few; some cities had the same scorer for all the games of a series, while others didn't; some scorers had extensive experience coaching or playing baseball, while others had none. Not surprisingly, she lamented, "Scoring is not only different city to city. It's different scorer to scorer."[4]

Given the centrality of statistics in the twenty-first century, it's perhaps surprising that the system of scoring was still just "good enough." Official scorers have received complaints since the first days of professional baseball, but this criticism has never been able to dislodge the system of ad hoc appointments, club involvement, and a fundamental lack of standardization. The system appears to have nevertheless succeeded in turning subjective judgments into objective facts.

It hasn't been easy, however. Across the history of official scoring a number of different mechanisms have been used to try to ensure objectivity, from restricting who scores to specifying how they score. For the first decades of baseball, official scorers were simply chosen by clubs and acted with authority as gentlemen and with data coordinated by the league. By the early 1900s, these club-affiliated gentlemen had given way to experienced baseball reporters. Over the following century, reporters were replaced by independent contractors who were appointed by the league and whose decisions were appealable and reviewable by the league. Throughout, bureaucratic control of scorers' judgments backed their authority. Official scorers have consistently managed to produce numbers through contestable judgments that, remarkably, have come to be treated as authoritative and objective statistics.

———

Before the advent of professional leagues, baseball's rules specified that there should be two people appointed to score the game, one from each club involved. Each side consequently kept its own score and reported it to newspapers, league offices, and other interested parties. The baseball reporter and scoring enthusiast Henry Chadwick noted that a scorer "should be one also whose gentlemanly conduct will render him acceptable to all who are liable to make inquiries of him relative to the score of the game." A "thorough gentleman" was needed in part because scorers were subjected to "endless" questioning from "incompetent reporters" and "interested spectators," which overall produced a "pretty severe trial of patience." Of course, he clarified, demeanor and good humor only went so far: the scorer should be "fond of statistical work and competent to make out a full analysis of the season's play each year."[5]

Scorers needed to be gentlemen because they were responsible for apportioning credit. The criterion of gentility had long been used in English philosophic and scientific traditions as a guarantor of reliability and commitment to truth. Gentlemen were presumed to be insulated from base interests and therefore resistant to the temptation to skew judgments for personal gain. A scorer, Chadwick explained, must "decide a point according to his honest and unprejudiced opinion." Chadwick gave hints for how to determine credit, noting that "in recording home runs," only those "out of the reach of the fielders" should count, whereas "home runs made from errors in the field . . . should not be counted." A scorer was supposed to "charge a catch as missed, if the ball touches the fielder's hands and he fails to hold it." Scorers were also to indicate when players were left on base after a clean hit, since "it frequently happens that a good hit fails to be rewarded with a run, from the fault of the striker following the one making the hit."[6] The scorer needed to be creditworthy because he was responsible for assigning credit and blame.

Official scoring may have been understood to require gentlemanly credibility, but experts were divided on whether anyone was capable of keeping score. Sometimes scoring was called "an easy proposition" that "can be mastered with little practice," whereas other times it was said that "it is no easy matter to score a ball game, play for play." Chadwick himself claimed that with his preprinted scorebooks in 1868, "any bright boy of ten years of age can record the particulars of a match game from which data can be obtained for a complete analysis of the batting and fielding of a game."[7] Yet Chadwick complained for many years that clubs were not keeping adequate scores, perhaps indicating it took more than a bright boy to manage the task.

By the 1860s, scorecards were proliferating among fans, who kept their own score as teams played. When Harvard's baseball club met the famously undefeated (and effectively professional) Cincinnati Red Stockings in June 1869, scorecards were printed for attendees, leaving space for fans to write in the outcome of each at-bat. The batting order and defensive positions were preprinted on the cards, meaning fans were forced to scratch off the listed player in the case of any last-minute substitution. While these cards were relatively primitive, with only basic columns to note outs and runs, fans at Harvard's games were keeping far more advanced score-cards by the 1870s, indicating the number of hits and bases made, how outs were recorded, and when players were left on base.[8]

The language of gentlemen made clear the presumption that scorers would be men. Yet in at least two cases, women early on served as official scorers. From 1882 to 1891 the official scorer for Chicago's National League club was Elisa Green Williams, who used the name E. G. Green to conceal her identity even from the league. Similarly, in South Carolina, Martha Boynton Reynolds was the official scorer for the club her husband managed and pitched for in 1905. There have not been more than a few other female official scorers in the century-plus since—though the rec-ommendation of hiring only gentlemen did not persist, scoring has remained largely a male domain.[9]

When the National League was formed in 1876, the scorer was responsible for forwarding a statement containing the "full score" of the game, its participants, and the date and place of its occur-rence within 24 hours to the league secretary. The secretary was charged with filing and compiling these reports and creating a "tabular statement" of games won and lost, a statement that would be the "sole evidence" determining the standings of teams in the league. Though no "official scorer" was labeled as such in this first year, there was no doubt that, as one reporter indicated before the season began, "it is made the duty of each club to appoint a proper man to score all games."[10] The rules did not specify this person had to have any special training or skills, just that he complete the league statistical form as the rules prescribed. Each club had to have its own scorer, but for league records, only the home club's scorer was asked to forward his score of the game.

Clubs found these instructions overly vague, and within a year, the league's playing rules further specified that "in order to promote uniformity," "scorers of League clubs" had to include in their report each batter's name, position, number of times at-bat, number of runs made, number of base-hits, and total bases reached off his hits, as well as those bases made by runners in front of him on his hits (this final provision did not last long). Each statistical summary was also to include a player's put-outs and errors made in the field. Forms were provided by the league office to standardize this process.[11] The scorer remained on the field, but scoring had symbolically and physically become a bureaucratic concern of the leagues rather than a recordkeeping practice of individual clubs.

The authority invested in the league explicitly acknowledged the role of character in the keeping of league statistics. As the league constitution specified by the 1880s, the secretary of the league—the person responsible for aggregating statistics—should be a "gentleman of intelligence, honesty, and good repute, who is versed in base ball matters, but who is not, in any manner, connected with the press, and who is not a member of any professional base ball club."[12] The secretary's credibility emerged from his independence as well as from his own character.

While the specific statistical categories scorers report have changed substantially over time, the basic practice of a scorer promptly reporting statistical information to the league office after each game has remained in place. This continuity in practice was supported, at least in the National League, by continuity in league office personnel. From 1871 to 1903, Nicholas Young was the secretary of the National Association of Professional Base Ball Players and the National League, as well as president of the National League during part of that period. In the secretarial role—and later as president—he was succeeded by John Heydler in 1903, who served into the 1930s. The statistics of the National League were thus handled for the first 40 years by Young and Heydler, and then from 1919 by the newly formed Elias Sports Bureau. Consequently, only three entities have been responsible for keeping official National League statistics since 1876.

By the early 1900s, scorers had to send their summaries to the league office immediately after each game (with a fine of $2 leveled against any scorer whose records were not at the league office within five days). Scorers used preprinted sheets, as prescribed by the rules, and judgments about hits and errors were considered final and could not be altered. The league secretary was originally allowed, however, to make certain calls himself, most notably the determination of the winning and losing pitchers of a game. The secretary was also ultimately in charge of compiling the records and sending out official averages in the fall after the close of the season.[13]

By 1913, Heydler apparently was running the office like the well-oiled calculating machine it was intended to be. As *Baseball Magazine* editor F. C. Lane detailed in a laudatory report of Heydler's recordkeeping, "Records have been systematically kept and jealously preserved since 1903" and "are complete in every detail to mathematical perfection." Heydler would accept the reports from scorers, paste them with newspaper clippings of unusual happenings into a scrapbook, and then personally enter the daily records of individual players and teams. First, he would add up columns to ensure that they logically checked out (e.g., team totals needed to match sums of individual performance), and then he would determine losing and winning pitchers. Then he would transfer the figures to a day-by-day or daily ledger, with 18 columns for various batting statistics and 17 for pitchers, thus 35 columns in all. Each new row would entail a new game, with home games entered in black and away games in red. Team records would go into a separate form and would again be checked against individual totals.[14] By 1900, the authority of official statistics resided firmly in the league office. That is, "official" statistics were by definition "of the office."

Heydler's ability to run his office effectively enabled the league to not worry as much about the individual scorers. Scorers operated as functionaries of the secretary and wouldn't even be listed in the rulebook or constitution as a representative or official of the league until midcentury. After all, like any bureaucrat, their authority emerged not from their own charisma or knowledge but from the office itself. The German sociologist (and contemporary of Heydler) Max Weber offered an account of "bureaucratic authority" that describes the National League statistical operation precisely: a clearly ordered system, centered on written documents handled according to established rules, that produces authoritative knowledge. In this case, that knowledge was of baseball statistics.[15]

Perhaps because leagues never specified exactly whom clubs should appoint as official scorers, there was often concern about possible discrepancies between official and unofficial scores. In 1887, Secretary Young wrote an open letter to scorers in *Spalding's Baseball Guide* called "Points on Scoring." He used the letter to note that the rules had been made more specific to try to answer the "large number of queries relative to the department of scoring." Remarking that "I do not think the high value of accuracy, care and impartiality in recording the points of the contest is fully appreciated," he also worried that the temptation to yield to the wishes of the local team was "frequently alluring." He compared the "official corps" of scorers with that of the newspapers, subject to "local pride" and boosterism. His letter was framed as a way of promoting more uniformity between newspaper scorers and official scorers by encouraging impartiality, especially as published statistics "go far to mould the public estimation of a player's value."[16]

Clubs themselves increasingly hired local newspaper reporters as official scorers. In this respect reporters had many virtues—they not only knew the rules of baseball but were often familiar with hometown players and front office personnel. They were already present at games keeping score, so additional expenses would be minimal. Reporters also needed league statistics for their accounts of the games, so they had a stake in ensuring credible statistics were kept.

As they became increasingly used as official scorers, reporters themselves sought to organize a professional association to manage access to their ranks. Reporters had tried previously to standardize baseball scoring and reporting by forming the Baseball Reporters' Association in 1887, but this and subsequent attempts had floundered.[17] By the first decade of the century, baseball writers were again pushing for a new organization to represent them and their interests.

The impetus for reform came during the 1908 World Series, when out-of-town writers were placed in the back row of the grandstand for the games in Chicago and "compelled to climb a ladder" to reach the roof of the first base pavilion for the games in Detroit. By the final game on October 14, a group of prominent writers had met in Detroit, and "a temporary working organization was effected." The first formal meeting of the Base Ball Writers' Association of America (BBWAA) followed that December in New York City, when Joseph Jackson of Detroit was elected president. The group's immediate concern was to "secure creature comfort" in parks by reducing the number of people allowed in press boxes and improving access to them for qualified writers. Both leagues—also meeting that December in New York—assured the newly formed BBWAA that they would cooperate to improve press box conditions and facilitate scoring reforms.[18]

Despite the group's claim that it was not attempting to monopolize baseball writing, the BBWAA was clearly attempting to limit who counted as a professional baseball writer. By policing the bounds of the profession, they could also police the ranks of official scorers. As one wire report put it, the goal was to "gain control of baseball press boxes." The group also promised to advocate on behalf of writers for changes to the game's rules and, at the request of the leagues, formed a committee to recommend a "more uniform system of scoring."[19]

Within a few years of forming, the new writers' association demonstrated its power to set standards. In 1913, the BBWAA's scoring committee advocated for a clearer stance on "earned runs," the limiting of assists to one per rundown, and the recording of catchers' success at throwing out prospective base stealers. They also adopted the "Cincinnati hit," replacing the fielder's choice ruling in situations in which both the runner on base and the batter-runner were ruled safe. Called this because its main advocate was Jack Ryder, a Cincinnati reporter and official scorer, the new rule was reversed by the BBWAA after a single season. The power of the organization to arbitrarily change scoring rules on a yearly basis led one *Baseball Magazine* writer to conclude in 1916 that "there was little system anywhere."[20]

The BBWAA did prove effective in its quest to regulate access to the press box. Soon after Kenesaw Mountain Landis became the first commissioner in 1921, the BBWAA became the sole source of scorers for the game. By the end of the 1930s, the organization's leadership could look back and note its success in improving press box conditions, simplifying the game's rules, promoting uniformity in scoring, and uniting North American baseball writers.[21]

The formation of the BBWAA alone didn't placate critics, mainly because scoring remained in the hands of hometown partisans. George Moreland in *Baseball Magazine* wrote of the need for a "better scoring system" in response to advances in strategy like the squeeze play, sacrifice hits, hit and run, and other "late ideas now practiced." What was more troublesome, in his view, was that scorers might adjust their rulings "simply to keep on the good side of such and such a paper."[22] Another *Baseball Magazine* reporter, M.V.B. Lyons, called the use of hometown writers "unbusiness-like" and noted the problem of decisions being made in secret by someone who was hired by the hometown club, employed by a hometown newspaper, and likely a resident of the town itself. The temptation of partiality, he contended, would be too great.[23]

Writers themselves tended to blame the leagues for the inconsistencies in official scoring. J. Ed Grillo, who served on the BBWAA's scoring committee, complained that despite the nascent group's attempts to promote standards, the president of the American League, Ban Johnson, continued to publish official averages that were based on incomplete data, rendering them "worthless and misleading." Before 1920, Johnson also had full authority to award wins and losses to pitchers without any consistent rules or explanation (the modern standard for awarding wins wouldn't appear in the rulebook until 1950). In marked contrast to Heydler's efficient National League operation, Grillo concluded in 1910 that "the American League reputation for issuing unreliable data of its playing seasons is now so well established that no one places the slightest reliance in any of its figures."[24] Writers like Grillo saw reliable data as the product of top-down coordination. Their emphasis was not on differing judgments among scorers but instead on effective bureaucratic management.

Concern with the management of individual scoring decisions was crystallized by two events, the Chalmers Award controversy of 1910 and the 1922 debacle over Ty Cobb's batting average. In 1910, the race for the batting title went down to the final day of the year, with Ty Cobb and Nap Lajoie competing not just for the American League's best batting average, but also for the automobile that the Chalmers Detroit Motor Company had promised the winner. As the season drew to a close, Americans around the country closely followed the box scores to see if Cobb would triumph over the aging fan favorite Lajoie. With Cobb skipping the final two games to try to preserve his slight advantage, Lajoie had a chance to overtake him. As Rick Huhn has recently examined in detail, Lajoie went eight for nine over the final doubleheader of the season, "earning" his hits under rather questionable circumstances. There was pressure placed on the official scorer (and bribes offered) to credit him with nine hits instead of eight—including by Lajoie himself, who called the official scorer after the games. The rulings on these hits would determine the batting champion. Because the race was so close, several reporters took the time to go back through the daily records, finding a number of uncertain official rulings over the course of the season. There had also been rumors circulating that Cobb was consistently given the benefit of the doubt by scorers, though there were clearly debatable calls on both players' records.[25] The race revealed that though the system did produce statistics, its inconsistencies and problems appeared substantial when placed under a microscope of any magnification.

American League president Ban Johnson held hearings, reviewed records, consulted the scorers and umpires of the final games, and ultimately determined that Cobb had narrowly edged out Lajoie for the title, .385 to .384. Chalmers, for its part, was happy to offer cars to both players, given the priceless publicity the campaign had earned for the company. And in a stranger-than-fiction coda, Pete Palmer, while going through the records as he compiled his statistical database nearly 70 years later, noted that one of Cobb's games had been mistakenly added into the record twice. Without the mistake, Lajoie would have finished with a higher average, though Major League Baseball, then and now, has refused to change the records.

After the 1910 season, league officials wanted to avoid any future controversies of this sort. The outrage of fans was directed not only at the fallible judgment of scorers but also at the appearance of conflicts of interest. National League president Thomas Lynch proposed to take away the clubs' power to appoint official scorers. One alternate possibility was for the BBWAA leadership to appoint the scorers for each city; another was to appoint "traveling" scorers not tied to any particular city. In fact, this was not the first time a "traveling" scorer idea had been floated in an effort to bring scoring in line with Lynch's former profession of umpiring. In a 1908 *Baseball Magazine* article, George Moreland had also advocated putting scoring on the same basis as umpiring, with scorers hired by the league, traveling from city to city.[26] Either reform would have served the purpose of preventing clubs from "playing favorites" with scorers.

The crucial issue was who controlled scorers' appointment. As the *Hartford Courant* complained in 1915, "The present system of having the club owners appoint the official scorers is the chief cause of the evils that have crept into the official records." Scorers wanted the money the extra job provided, and owners had incentive to replace unfriendly scorers. Given the proliferation of questionable rulings in both the minor and major leagues, it was safe to infer that "the scorer who tries to be honest and score consistently stands small chance of being appointed official scorer in some cities." The *Courant* editorial concluded that the only ways to eliminate this easy "graft" were to give the BBWAA control over appointing scorers without pay or to have the league presidents appoint scorers directly without providing their names (a system the National League had used briefly in the nineteenth century).[27] There was little agreement, however, on whether scorers would be more reliable if they were known or if they remained anonymous. In the end no good fix was found, and the system remained as it was.

The ad hoc agreement between clubs and local reporters again came under acute pressure in 1922 as the ever-controversial Ty Cobb flirted with finishing the season with a batting average of .400. One hit would make the difference, so again intense scrutiny was applied retrospectively to every scorer's decision. Months earlier, on May 15, the official scorer had submitted a statistical summary giving Cobb one hit and indicating he also reached on a fielding error made during this rainy game. The American League's official statistician, Irwin Howe, however, had apparently chosen to rely on the scoresheet kept by the Associated Press scorer for that particular game. The AP scorer had given Cobb two hits. Howe justified his decision to use the AP score by noting that the official scoresheet suffered from a number of errors, lacked the signature of the scorer, omitted the name of a pitcher in the summary, and overlooked a triple that had occurred elsewhere in the game. Accepting the AP score as more reliable, Howe made it the official score of the game, claiming "this same procedure" had been "followed in many previous cases."[28] His decision only became controversial after the season because the "extra" hit put Cobb over .400.

Scorers from New York—the site of the disputed hit—protested against the "arbitrary" decision of Commissioner Landis and President Johnson to support Howe's stance.[29] They not only questioned the point of having a scorer deemed "official" if it was possible to overrule him but also worried that such actions undercut the autonomy of the scorers themselves. As with the 1910 game, the issue was twofold. On one hand, the dispute concerned whether the official score was a product solely of the judgment of the person appointed or a combination of his judgment and that of the league's front office. On the other hand, it concerned how to manage scorers and deal with possible problems of judgment and execution. The New York scorers brought the matter to the BBWAA, which narrowly supported them. Nevertheless, the league maintained its stance, the baseball guides followed, and to this day Cobb officially enjoys a .401 average for 1922.

Complicating the complaint of the New York writers was the fact that when Howe chose to use the AP score over the official scorer's report, he was choosing the scoring summary made by the president of the BBWAA, Fred Lieb, a major figure in early-twentieth-century baseball reporting. (The official scorer was John Kieran of the *New York Tribune*, who was far less experienced.) By relying on Lieb's score, Howe, Johnson, and Landis were deferring to expert judgment over nominally official status (there were also rumors that Kieran wasn't actually in the press box when the disputed hit occurred). The controversy wasn't just about "getting it right"; it was also about the legitimacy of the system, especially given the fact that the same person could be an official scorer one day and an unofficial one the next. Lieb *himself* advocated for Kieran's score, not his own, to preserve the authority of the BBWAA's selection of scorers.[30] The weaker the boundary between unofficial and official statistics, the more it needed to be respected for scorers to have authority, reporters claimed. The league responded that with such a weak boundary, it had the responsibility to use the scoring summary that was the most reliable rather than whichever happened to have been designated official that day. In response, the New York members of the BBWAA refused for a time in 1923 to serve as official scorers, though eventually they came around.

For President Johnson the whole affair was an opportunity to publicly lament what he saw as the sorry state of scoring. He complained in a letter to the BBWAA that "this office has no direct dealings with the official scorers," who are "appointed by the clubs in various cities." "Experience has plainly shown," he continued, "that in many cases they are grossly lacking in efficiency and responsibility." At various times, he noted, the league office had had difficulty receiving official scores consistently, and at one point, the league had had to appoint an alternate scorer and pay him directly because the club's scorer failed to send in any reports. Johnson suspected that the problem was the cozy relationship BBWAA members enjoyed with clubs: with half a dozen or so local reporters supplementing their income with scoring duties, it seemed possible that this favoritism had unconsciously affected them. He concluded that "it would seem the part of wisdom and prudence for the baseball writers to put their house in order before sending me scurrilous and questionable complaints." The next spring, Johnson continued to warn reporters that the leagues could take back control of scoring from them and appoint official scorers as they did umpires, a largely empty threat given the added expense of the proposition and the desires of the club owners to maintain control over reporters' as well as players' statistics.[31]

Though changing little directly, these controversies did have an eventual effect. Scorers began to announce their decisions publicly in the press box soon after plays rather than keeping them private. And by 1929, the league presidents had convinced the club owners to allow the leagues to appoint scorers directly, on the basis of the BBWAA's recommendation of its own members. The owners also agreed to have only one official scorer and two alternates appointed for each city.[32] The decision to continue drawing directly from local BBWAA members was clearly a cost-saving measure. Through the 1930s and 1940s scorers received about $13 per game for filling out and sending in the summary sheet. Since they were already keeping score for their day job, what would have been a meager amount for an outsider remained a nice little bonus for them.

Not everyone was completely happy with the control the BBWAA retained. In 1949, Ford Frick—then National League president, but soon to be appointed commissioner—gave an extended interview to the *Sporting News* to complain about the declining standards of scoring. Since the 1929 agreement, it had become de facto practice for the leagues to accept the list of scorers the BBWAA recommended. Frick noted, however, that the BBWAA had gradually moved away from the practice of appointing one or two scorers per city and had again shared the scoring duties among many reporters, not all of whom were qualified.[33] In response, writers complained that the pay for scoring was so pathetic that they had trouble ensuring that a small number of experienced reporters could cover every game. Frick listened; scorers' pay did increase over the 1950s to closer to $20 per game. For its part, the BBWAA also formally adopted the requirement that members had to cover at least 100 games in each of five seasons (later reduced to three seasons) to be eligible to officially score.

These measures, minor as they might have been, were clearly intended to boost the credibility of the statistics. The BBWAA promised to appoint only experienced, knowledgeable reporters as official scorers. The league promised to treat them as professionals and not override the official scorer's account with one from a different reporter. With Frick's support, official scorers also received their first formal recognition. When the official rules were reorganized and rewritten in the late 1940s, one of the achievements of the Scoring Rules Committee (consisting of the

BBWAA's Roscoe McGowen, Dan Daniel, Halsey Hall, Charles Young, and Tom Swope) was ensuring that the new rules explicitly specified that "the scorer is an actual official of the game he is scoring, is an accredited representative of the League, is entitled to the respect and dignity of his office and shall be accorded full protection by the President of the League." In 1955, the rules clarified that official scorers had "sole authority" to make decisions involving judgment. And in 1957, they were revised to indicate that scorers were appointed by the league presidents and were to be seated in the press box.[34] These new guidelines helped formalize the informal agreements of the early part of the century. By and large, the efforts worked: reporters, appointed by the BBWAA and paid by the league, were able to score games in such a way that their judgments were taken as authoritative.

By the 1970s, however, there were a number of interconnected forces putting pressure on this system. First, newspapers were increasingly reluctant to allow reporters to act as official scorers for games they were covering. Starting with the *Washington Post* and then gradually extending to newspapers in New York, Milwaukee, Pittsburgh, and San Francisco, paper owners prohibited their reporters from also serving as official scorers. Reasons varied, but they included the fear that reporters would hamper their ability to get clubhouse access if they made controversial scoring decisions that went against the team's interest. Furthermore, some writers noted that BBWAA chapter meetings had devolved into petty politics over the selection of scorers, with some irrational results, like Boston's appointment of nine official scorers, none of whom were among the city's most experienced baseball reporters.[35]

The increased visibility of scorers in this period also made the press nervous. Since the nineteenth century, the idea of anonymous scorers had had defenders, most notably Nicholas Young, who thought that their identity "ought to remain a secret as far as possible." Similarly, during the debates after the founding of the BBWAA, Thomas Lynch came out in favor of keeping the scorer

and the scores secret until the official averages were released at
the end of the year. By midcentury, the scorer was still largely
anonymous, but his calls were ever less so, with new electronic
scoreboards flashing "hit" or "error" immediately after calls, and
radio announcers debating whether a hit should have been ruled
an error. One reporter lamented in 1949 that it wouldn't be long
before his name was added to the scoreboard or program like the
umpires'. Sure enough, by the 1960s, the BBWAA had agreed to
allow scorers to be identified publicly before games.[36] It was no
longer possible to look the other way and pretend that a particular
reporter was simply not the official scorer that day—if a ruling
rankled a player or manager, he could take out his anger directly
against that reporter and his paper.

The conflicts between newspapers and players were height-
ened because players were increasingly concerned with their own
statistics. Players and owners had used statistics to argue for pay
cuts and raises since the first professional games, but by the 1970s,
the stakes had become much higher. In 1976, the National League
approved the creation of a contract for Tom Seaver that based his
pay in part on various statistical measures. These "performance
contracts" would become increasingly popular in the era of free
agency. New negotiations over the status and rights of free agents
required accurate official batting, fielding, and pitching statistics.[37]
Players' wishes for salaries commensurate with performance required a reliable measure of performance, putting official scorers'
decisions in the spotlight and creating possible conflicts for reporters who created the statistics and then reported on the salary
negotiations based on them.

Some writers continued to defend the practice of BBWAA
members' monopoly on scoring, none more so than Jack Lang.
A longtime local and national leader of the BBWAA, Lang served
either as secretary, assistant secretary, or secretary emeritus of
the organization from 1966 to 2001. Starting in 1946 with the *Long
Island Press* and continuing with the *New York Daily News*, Lang
regularly defended the integrity of reporters as official scorers

(and, not incidentally, as voters for Hall of Fame elections). As more papers defected from this position, Lang continued to assert the value of reporters as scorers. As he noted repeatedly, ball clubs didn't get to pick scorers, and they only had the "smallest particle of say" in assigning scorers to games. The BBWAA held primary responsibility. The pay came right from the league office, so there was no question of direct conflict between the press and clubs. Furthermore, he contended, the BBWAA's eligibility rule of covering 100 major league games over at least three seasons ensured that scorers were experts at the game, and given the low pay (raised to $50 by 1976), there simply weren't more capable scorers available. League representatives generally agreed, noting in the late 1970s that hiring a fifth umpire or a retired player would cost them many times the current outlay for scorers. Commissioner Bowie Kuhn noted in 1978 that he wasn't "pushing for change" because "the current system seems to work reasonably well."[38]

Nevertheless, even Lang had to admit defeat by 1979, and the BBWAA voted at its annual meeting that December to inform the leagues that it could no longer be responsible for providing all of their official scorers. Given the controversy, league officials had already been preparing to provide scorers in cities where there were no BBWAA-qualified scorers available. The BBWAA took a straw poll to gauge interest in removing members from scoring altogether but found that sentiment was "not overwhelmingly against" the use of reporters, so those whose employers allowed them to act as official scorers were permitted to continue. By the 1980 season, however, 15 official scorers were not members of the BBWAA, and while some were retired writers, others had day jobs as real estate agents, construction workers, drivers' education teachers, and salesmen.[39]

The justification that reporters could be effective scorers by virtue of their experience and proximity to the ballpark could not ultimately withstand the pressure from newspaper owners, though the leagues themselves seemed content to continue this convenient arrangement. There was widespread suspicion among officials that it might not be easy to find replacement scorers who would be sufficiently knowledgeable, acceptable to club ownership, and willing to work cheaply. They were right: after the BBWAA ended its agreement with the leagues in 1979, the leagues claimed they had difficulty finding qualified people to score. More than one reporter noted with glee that there were so many complaints about scoring in the first two seasons after the change that the American League sent a directive to clubs in July 1982 asking for suggestions for improvement.[40] The ending of the agreement with the BBWAA wasn't evident anywhere in the leagues' rules—officially, league presidents still appointed scorers, who carried on as before—but it meant that scorers no longer had the legitimacy provided by an affiliation with seasoned, professional baseball writers.

———

The selection of scorers—the determination of who pays them and who hires them—was not the only way to turn subjective judgments into reliable statistics. In fact, there has been a long tradition of attempting to eliminate bias and create reliability through written rules. Scorebooks as early as the 1870s had instructions for how to notate hits, outs, and runs, but rules for official scoring were not really about how to mark a scoresheet. Official scorers needed to learn how to categorize certain actions, distinguishing hits from errors, passed balls from wild pitches, and earned from unearned runs in an acceptable way. They needed to learn to discipline their judgment in order to classify events correctly.

Angry readers were writing letters to newspaper editors on the matter of inconsistent scoring as early as 1873, when each team was still providing its own scorer. In one letter to the *New York Clipper* the writer noted that in a crucial game, with two outs and men on base, a "hot line ball" was hit to the third basemen, and, "not going fairly to him," "went through his hands." The scorer for the batsman's club (and the reporters present) credited him with a base hit, while the other club's scorer gave an error to the fielder. Because three or four runs were scored after this play, the decision made a difference in the crediting of "earned runs" to the pitcher. Henry Chadwick himself would respond to the disagreement the next year in an article, "Base-Hits and Earned Runs." "With all due regard for the capabilities of the many intelligent and competent scorers who have had to decide upon the questions of base-hits scored and runs earned during the base-ball campaign of 1873," he opened, "it is a fact well known that the majority are so involuntarily biased by their connection with the clubs for which they score, that scarcely any two can be found who are in accord." In cases like this, he declared, "when 'doctors disagree,' an outside individual, who is removed from party bias, must step in and decide the disputed point"—and that was precisely what he intended to do. Guided "solely by our efforts to promote the best interests of the game in bringing it up to the highest standpoint of a scientific field-sport," Chadwick gave one of the first sets of guidelines for scorers on rulings concerning hits and errors, including his favored method of "quoting from actual play."[41] Chadwick enumerated the various cases to try to discipline scorers' judgment in a way consistent with his vision of baseball as a "scientific field-sport." According to him, there were four cases of obvious hits—essentially when balls are hit out of reach—and two cases in which some difficulty in judgment might occur. He advocated deciding the matter on the basis of what was *normal* and what was *exceptional*. In cases in which the normal limits of the ability of fielders were "exceeded," hits rather than errors should be indicated. His experience enabled him to determine what counted as normal or exceptional and therefore to "decide the disputed point." Scorers, by implication, needed similar experience to do the same.

Not surprisingly, when the National League started play a couple years later, Chadwick suggested in the *Clipper* that it would be "advisable" for the convention of league players to "add to their rules a brief code of instructions for their club-scorers, containing explicit rules for recording" hits and errors. As things stood, the paper warned, scorers disagreed on what counted as a base hit or a misplay, and when scorers sent their records to the league secretary at the end of the season, "scarcely two records out of eight are fair or correct." Indeed, in the next year's official rules, the National League did attempt to explain what constituted a hit, defining it as occurring in cases when a fair ball was hit "out of the reach" of fielders, hit so that it was partially or wholly stopped by a player in motion who was unable to recover in time, or hit so sharply or so slowly that the fielder could not handle it in time. Similar rules applied for scoring assists, put-outs, and errors. The rules added, though, that "in case of doubt," the scorer was to "score a base hit and exempt the fielder from the charge of error."[42] The league seemed to have less faith than Chadwick in scorers' ability to discern what was exceptional and so made it clear that an error should be given only in doubt-free situations.

The many subsequent attempts to clarify the matter suggests that the league's description wasn't quite adequate. T. H. Murnane's popular *How to Play Baseball* series added some "rules of thumb" for scorers, clarifying cases of short hops or rough surfaces (hit, not error) as well as those when a fielder had two hands on the ball at some point (error, not hit). In 1911, J. M. Cummings, an official scorer for Baltimore, published his *Practical Textbook* on scoring as part of Spalding's Athletic Library. Clearly intended for newer scorers, he noted that the rules left "too much to be interpreted by the beginner." As a result, not all scorers had "acted in unison, forming their opinions from some fixed, acknowledged standard." Instead, they had, in his metaphor, erected different "architectures" or "superstructures" on the foundation of the rules. Honesty, good judgment, and a clean conscience were essential to any scorer, but accuracy in all things was also required, and yet surprisingly difficult to achieve.[43]

Scoring, for Cummings, ideally entailed people acting like machines. Scorers had to know how to apply the rules and general principles, needed to maintain "eternal vigilance" in following the ball, and could never be in a position to benefit from a decision.

Perhaps most curiously, a scorer was tasked with performing his duties "impartially[,] as though the players were inanimate objects he had never before seen and he never expected to see again." This trait Cummings considered essential for what was perhaps the "most intricate thing" required of scorers: distinguishing hits from errors. Again, what was required was not knowledge of the individuals involved, but an understanding of "reasonable expectation"—if the fielder could not be reasonably expected to make the out or if "manifestly phenomenal" work was required, then a "hit" should be credited to the batter.[44]

The following year, I. E. Sanborn—who with Murnane had been among the original founders of the BBWAA—followed with a guide on "How to Score a Baseball Game." He lamented that despite efforts to "produce uniformity" since the formation of the BBWAA, "no two authorities will agree on every point in a game." Returning to the language of credit and credibility, Sanborn noted that the "chief difference" between such authorities was in the "crediting of hits and the charging of errors." Since this was "considerably a matter of individual judgment," he considered that "perfect uniformity cannot be hoped for," and "no hard and fast rule can be made to fit such cases." Nevertheless, he believed that it was helpful in such situations to try to determine whether the play might have been made by "the exercise of ordinary skill."[45] Sanborn replaced Cummings's language of "reasonable expectation" with that of judgment of the "ordinary"—substituting the language of experience for that of rationality.

In 1920, the National League found it necessary to clarify that official scoring required scorers' "best judgment." In cases of doubt, scorers were told to consult league headquarters, though they should remember that in some cases, "it is not possible to make hard and fast rules."[46] From the league's perspective, there were no clear rules, but rather an opportunity for deliberative judgment.

Attempts to manufacture uniformity and objectivity through guidelines continued unabated. In 1925, H. G. Fisher attempted to provide his own guidelines in his volume *How to Score*, part of his "Baseball Decisions" series. Concerned less with the sort of basic judgment on which Sanborn and Cummings focused, Fisher centered his book on situations that called upon deep knowledge of the rules. Organized as a series of questions and answers, the book primed scorers to understand how the various scoring rules worked in tandem. Fifteen years later, the minor leagues' "Official Scorers Bureau" published a manual for league scorers to try to create some uniformity among the hundreds of people scoring games at this level. A scoring newsletter was also briefly self-published in the late 1960s by Wirt Gammon, Sr., intended as an informal way for scorers to create uniformity in dealing with tricky situations.[47]

One strategy for uniformity was to defer to "common" sense. The *Sporting News Guide*, for example, noted that "in scoring always exercise common sense." Similarly, Dick Young, an official scorer in New York, noted in 1971 that while he couldn't say which way was the "best way" to score, "blame" was the "essence of scoring." Too often, he maintained, scorers were "slaves" to the written rules. "Common sense is a better servant," he said. "Common sense serves the true duty of the official scorer, for what is his purpose? To reward the deserving; to blame those responsible." Nevertheless, since scoring a ball game "is in the eyes of the beholder," it matters who is scoring the game and how he or she attributes the blame for a play.[48]

At issue here were not just different visions of the job—the scorer as impartial judge or the scorer as mechanical robot. Also in contention was how the duties of the official scorer might best be carried out by those who were doing the scoring—that is, newspaper reporters. Former player Murnane's *How to Play Baseball* series from the first years of the twentieth century had been suspicious that as the official scorers transitioned from club members who had played baseball to those who had only watched the game, it would change the nature of the scorers' judgment. Scorers who didn't play the game but rather learned "by book and observation," Murnane claimed, tended to be too hard on players and "ever anxious to credit errors."[49] He suggested that guidelines based on observational, not experiential, knowledge would be the best way to shape scorers' decisions. Newspapermen, after all, knew the ordinary because they had seen so many games. The most successful instructions for how to deploy judgment consequently relied upon Sanborn's expression of "ordinary effort." The phrase itself was eventually included in the guidelines for official scorers when the rulebook was overhauled in 1949, certifying the link between judgment and repeated observation.

Though there have been numerous tweaks to the scoring regulations over the years, the basic premise persists. The league presidents (or, more recently, the commissioner's office) appoint scorers, who have "sole authority to make all decisions involving judgment." (The scorer is of course not allowed to contradict either the rules or an umpire's decision.) The language of moral accounting remains ("credit the batter"; "charged with an error"; "credit participation"; "an earned run is a run for which the pitcher is held accountable"), as does the emphasis on basing judgment in observational abilities ("ordinary effort"; "exceptionally good fielding"; "unnatural bounce").[50] Scorers themselves consider their own objectivity to be essential. By this they don't mean that they'll always agree with one another or with the players involved, but rather that they will strive to judge honestly, consistently, and fairly. One longtime scorer, Ed Munson, said he ensured fairness by asking himself before assigning an error whether he'd have given it if his mother had been the one involved. Munson's version of objectivity involved treating people consistently rather than thinking about whether he knew a player or cared about a team.[51] For the same reason, scorers are discouraged from participating in fantasy leagues or any kind of wagering on outcomes (even outcomes not under their control like wins and losses). In this view, the authority to make judgments about credit is never separable from the moral credibility of the person scoring.

Official scorers are now considered full representatives of the league, though they no longer enjoy the credibility that membership in the BBWAA provided. Their authority comes solely from their position. This is true in a figurative sense—they are backed by the league because the league hires and fires them. But it is also true in a physical sense—the rules specify the scorer must take a position in the press box. This position is by definition a privileged one: no one else is sitting in the scorer's place, with his or her vantage point, and those watching from home, or from the field itself, can't claim to have seen exactly what the official scorer saw. (MLB scorers do have access to a system in the press box that enables them to easily watch video of plays they'd like to see again.) Though there's no formal guarantee of their expertise, by serving as representative of the league, scorers are "entitled" to the same respect and dignity afforded to the league office. As they sit in the press box watching the game, they embody the league's judgment about credit.

————

By the 1980s, there were rumblings for change from a relatively new source: players. Players, and free agents in particular, had been driving a focus on accurate statistics and unbiased official scorers since the first free-agent contracts in the 1970s. As part of the resolution of baseball's 1981 strike, players and owners agreed on a system of compensation for free agents that was dependent upon statistics. Starting with the 1981 off-season, teams losing "type A" free agents, defined as players in the top 20 percent statistically at their position, or "type B" free agents, those between the twentieth and thirtieth percentiles, would be compensated with additional draft choices.

On November 8, 1981, the *Chicago Tribune*'s Jerome Holtzman announced that as a result of the agreement, the top players had been ranked by position, with the rankings published for the first time.[52] Each player was given a point total, with 100 being top in every statistical category and scores then moving down from there (the top score in 1981, 98.925, belonged to pitcher Steve Carlton, who led in four of the six statistical categories for pitchers). The rankings were based on the previous two years' totals, or the imputed totals if players had spent time on the disabled list. Though the numbers had been calculated secretly by the Elias Sports Bureau, they were published for all to see. Scorers' decisions were suddenly even more visible.

Not surprisingly, some commentators, Holtzman included, immediately complained about the omissions, declaring that statistics failed to measure competitive drive or a player's enthusiasm or defensive contribution. Nevertheless, the rankings had power simply by their publication: now it was clear that of the 28 first basemen ranked for 1981's off-season, Cecil Cooper led the American League with a 95.926 score, while Greg Wells scraped the bottom with a paltry 14.444. Both teams and players had significant incentives to pay attention to statistics. Because the measures used to determine the offensive value of a player were plate appearances, batting average, on-base percentage, home runs, and runs batted in, every hit suddenly made a difference. The classification of players into statistically determined types continued over the 1980s, and statistics were also increasingly used in salary arbitration cases. (By 2012, free-agent types had been replaced by the use of "qualifying offers" in the basic agreement.)

The prominence of statistical performance and the emphasis on making ever-finer calculations of overall ability spurred players' interest in the quality of official scorers. By 2001, MLB had appointed a five-person committee to review official scorers' calls, and though the existence of the committee appeared nowhere in the official rules, appeals had doubled to about one a week by 2003.[53] By 2007, the appeals process had been formalized, and the following year's rules clarified that players or clubs could request a review of the official scorer's judgment calls. As had been the case for decades, scorers had 24 hours after the game to finalize their decisions and then submit them on a prescribed form to the league office (or directly to the official league statistician). Appealing clubs and players had 24 hours after the decision was submitted to notify the commissioner's office of their intent to appeal and an additional two business days to submit any testimony or video evidence in support of their appeal. Three outcomes were possible: the office could demand that a scorer change his or her judgment call if the judgment was "clearly erroneous," the office could "request" that the scorer change his or her call if the evidence warranted it, or the office could uphold the original call and impose a "reasonable fee" on the requesting party. No judgment decision would be changed after this appeal.

The process was streamlined again in 2012, explicitly putting the executive vice president for baseball operations—then Joe Torre—in charge of all appeals and giving teams 72 hours to appeal decisions after the official scoring summaries were submitted. The executive vice president was granted access to "all relevant and available video" and any other evidence he "wishes to consider." There was no longer need for a committee, although one could be used if the executive vice president desired. Torre used a committee to help him make a decision only once in his first three seasons as the final adjudicator of scoring appeals, when R. A. Dickey's one-hit shutout in 2012 was appealed on the basis that the only hit should have been ruled an error. The official scorer's ruling was confirmed. Though in theory any play, even one deciding a no-hitter, could be appealed, there was little incentive to retroactively award no-hitters from the commissioner's office.

The 2012 system of appeals—as specified in the official rules and agreed upon in collective bargaining—put the entire matter in the hands of the executive vice president. Club officials and players were explicitly prohibited from communicating with the official scorer regarding his judgments—because any of them could appeal, there was no need to confront scorers directly. The executive vice president could, with no explanation, no consultation of others, and no time limit, simply order a change in scoring when the scorer's judgment was appealed and deemed faulty. He did not receive all the appeals directly, of course; others in the office initially screened them to ensure frivolous ones didn't take up too much time. The number of appeals surged under the new system, from about 30 in 2003 to about 60 in 2011 to over 400 in the 2015 season.[54]

The appeals system certainly didn't eliminate the need for credibility. It simply put the onus entirely on the executive vice president for baseball operations. Merhige herself explained the process's initial success by pointing to how players and teams trusted Torre's judgment because of their perception of him as a "guy of high integrity. He tells the truth; he wants to get it right." He was a good choice because he was "highly respected" and was "considered an expert on field matters: he's played, he's managed, he's coached." As Stew Thornley, one of the official scorers who worked closely with baseball's front office, explained, "Joe is as respected as anybody can be in baseball," both from his background as a player and manager, and "beyond that just for his character and fairness." Scorer Ed Munson called Torre "the Chief Justice."[55] This credibility was essential in part because what Torre was ultimately deciding in the case of, say, a hit versus an error was whether the play was made with "ordinary" effort. When he changed an error into a hit, he was effectively ruling that the scorer's judgment of "ordinariness" had been erroneous—that he or she had failed to recognize what counted as an ordinary play.

Predictably, not all scorers were happy with the fact that any one of their judgments could be appealed simply on the whim of an unhappy player. Not that Torre overturned a lot in his first three years: less than 25 percent of all appeals in 2014 and 15 percent in 2015 were overturned, compared with nearly 50 percent of all appeals of umpires' rulings each year (though a factor in the latter may have been the higher cost of appealing an umpire's decision, since such appeals were limited). One clear advantage of this system, though, as Thornley pointed out, was that the "stress level in the press box has been reduced by a huge amount," since any player, manager, or team could just go straight to the commissioner's office.[56]

After the 2012 revision, appeals could involve cases only of judgment, not of interpretation. Official scorers could not violate either umpires' decisions or the rules themselves, and when they did, the commissioner's office could just make the necessary change directly. In practice, what this meant was that if an Elias Sports Bureau representative noticed an official scorer had marked a "save" for a pitcher in a situation in which he was ineligible for one, the representative would contact the commissioner's office, which would notify all involved and have the scorer make the change. There was a clear line between the application of judgment—the domain of the scorer, subject to appeal to the executive vice president for baseball operations—and the application of the rules themselves, which was the domain of the commissioner's office, through consultation with the Elias Sports Bureau.

The appeals process was a solution to the problem of increasing attention to statistics that did not require radical changes in the hiring or evaluation of scorers. Implicitly, all scorers' judgments were backed not just by the ability to "take another look" at a play on video, but also by the authority of Major League Baseball's front office. As independent contractors, hired and fired at the whim of the league, "official" scorers acted as an appendage of the front "office" rather than as individuals with their own independent authority or credibility. That's in part why almost no one knows who scores hometown games: that would imply that *who* these scorers are matters, when the goal of the system is simply to portray them as nameless but competent bureaucratic officials.

———

The present system of official scoring is obviously rooted in the 1870s. The scorer is appointed in part on the basis of his or her familiarity with the club; makes judgment calls pertaining to the determination of hits, earned runs, and errors; records the league-mandated statistics for the game; and is responsible for submitting the summary to the league office within a fixed time period after each game. Complaints, then and now, revolve around the issues of uniformity, competence, bias, and expertise. Standardizing judgment, the central dilemma for Chadwick, remains the focus of MLB's management of official scorers.

During her tenure, Merhige, unlike Chadwick, did not try to create standardization solely through written instructions. Rather, she had the idea to create it by bringing scorers into the same room. Before the twenty-first century, official scorers had no forum in which they could share knowledge or experience, aside from meetings of the BBWAA. Each operated in his or her own world. That's part of the reason why the major controversies—like that of Ty Cobb's batting average—were so jarring: they pointed out that statistics, unlike scorers, were not local.

Merhige changed this. To "standardize," she explained, "it was my vision that they needed to get together in a room." So in 2011 she inaugurated the "Official Scorers' Seminar," an annual gathering of official scorers from each city to discuss plays and situations, trade experiences and tips, and create a sense of camaraderie. She also oversaw the creation of an online bulletin board where the scorers could congregate to post plays, ask the group questions, or simply pass on relevant information. The point was to create a repository of knowledge, of experience, that previously might have been hard to collect. If uniform written rules hadn't done the trick, perhaps bringing people together would.

The task turned out to be harder than expected, however. For one of the first meetings, Merhige had the idea to prepare a video with clips of plays that were difficult to score and then ask the assembled scorers to make the call themselves and discuss the correct call with three veteran scorers. But when the veteran scorers saw the video in preparation for the meeting, they couldn't agree on a single call. The video was scrapped.

This is one lesson of the history of scoring: Judgment about what's ordinary is rarely common. Every call is different. Such determinations are a matter of which differences matter, and the Official Scorers' Seminar—as well as a casebook that was organized in conjunction with it—was organized around typical scenarios: whether balls in the dirt should be called wild pitches, whether balls lost in the lights should be marked as hits, whether balls tak-ing short hops or falling between multiple players should be scored as errors. Tricky and rare cases abound, but instead of letting each scorer discover them on his or her own, it is easier to try to create a repository of collected wisdom. Is a given play a standard "ball in the lights" scenario, or are there relevant circumstances that differ from the usual case? The art of scoring, as some scorers refer to it, is not so much about passing judgment on plays as if each were unique as it is about being able to connect a particular play to the right category.

Since scorers can and do disagree, Merhige and others behind the Official Scorers' Seminar thought it better to try to minimize those cases by collectively agreeing upon as much as possible. Some scorers will always be prone to giving hits, and others to giving errors: even though one scorer from San Francisco may defer to batters, one in Washington, DC, may have the instinct to mark an error and then see if he can be convinced to make it a hit. Perhaps things are, as Thornley claimed, just inescapably "subjective." In this view, there "isn't a right call," only a particular scorer's call.[57]

Questions of imagination also come into play. Many appeals, after all, involve the determination of earned runs, and particu-larly runs that would have been scored had fielders been com-petent. While some situations can be covered explicitly by the rules, others require creative or imaginative determinations of what "would have happened" had the error not been committed. The best one can hope for, then, is to maximize consistency, both across one scorer's work and from scorer to scorer, acknowledging that there will always be some disputable cases.

Under the appeals process, scorers who faced appeals of their rulings were aware that if their judgment differed from the execu-tive vice president's, they risked being overruled. Consequently, his judgment became, at least for some, a de facto standard for determining what was "ordinary" or "usual." The appeals process treated official scorers' work as a first draft that could then be appealed in important or tricky cases and adjusted or corrected by the "expert." In the past, Torre attended part of the yearly scorers' seminar to both reassure scorers that they were on the same team (one scorer recalled it was "almost like a family") and reassert the executive vice president's standing as the final arbiter. The appeals system certainly wasn't democratic: even in some cases where nearly every scorer present agreed on a call, Torre stepped in to explain why the consensus decision was wrong.[58] Judgments about hits and earned runs were made by a particular scorer at a particular time, but that scorer also tried to predict what the executive vice president would see on video replay if the decision were appealed. If a scorer's judgment proved too inconsistent or fallible, he or she could always be replaced.

Ultimately, a scorer's job is different than that of an umpire; whether or not a player touches a base before the ball arrives is a claim distinct from that of whether the play might have been made with ordinary effort. It might depend on which scorer is in the press box that day, what he or she sees initially, how he or she is feeling, what he or she ate that morning, or whether the game is exciting. We might imagine that calls of ball or strike, out or safe, could, in theory, be made without human involvement. No one pretends the same is true of a call of hit or error. The scorer's judgment is inescapably tied to the scorer's subjectivity, the reality of his or her body existing at a particular place and time.

Because of this, Major League Baseball hasn't tried to eliminate scorers' judgment but rather has marshaled it into the service of objective facts. MLB has thus tried to turn subjectivities into objectivities. The ways it has done that are a veritable list of different conceptions of objectivity.[59] Objectivity might emerge from disinterested judgments. Indeed, the emphasis on gentility in the early years of baseball scoring was an effort of precisely this kind, and even today the appearance of impartiality remains so important that scorers' participation in fantasy leagues or other possible conflicts of interest is discouraged if not outright forbidden. Objectivity might emerge from trained judgments. Scorers once had to take an "exam" on the baseball rules, and even after that practice ended, they endured a form of apprenticeship, with current scorers and club officials evaluating their work in a limited setting before they were given full authority. Objectivity might emerge from mechanization, and some, like J. M. Cummings, have emphasized the ways in which scorers should try to emulate machines. Objectivity might emerge from rules, and over time the rulebook has been amended to specify, and clarify, the appropriate judgment in previously ambiguous situations. Objectivity might emerge from intersubjectivity, or consensus. That's precisely what the Official Scorers' Seminar was meant to foster. What is clear is that objectivity and subjectivity are neither "pure" categories nor easily distinguishable in every case, and the process of turning judgments into statistical facts has not smoothly transitioned from one regime to another, any more than the practice has obviously become more objective over time. It's a messy business.

One consistent approach, however, has been to emphasize bureaucratic objectivity. As Merhige lamented, after the scorers were no longer drawn from the self-policed ranks of the BBWAA, "the biggest problem these guys have is that they don't have a credential. . . . These guys absolutely know what they're doing but . . . because they don't have a title, it's hard to convince anyone that they're as good as I think they are." Indeed, early scorers were closely connected to clubs and represented a club's own record of the game. Eventually, official scorers came to report to the league, and the authority of their judgment was grounded not in their own personal charisma or ability, but rather in the secretary of the league. Likewise, the assignment of scorers vetted by the BBWAA ensured a certain level of credibility. As *Baseball Magazine*'s editorial board wrote just a few years after the BBWAA's creation, "The Association has given these writers a certain prestige, which invariably results from close-knit organization and there is no doubt it will increase in well merited influence as the years pass by."[60] Now the basis of official scorers' authority has shifted to the executive vice president for baseball operations and other front office staff. And while Torre clearly commanded respect when he became the first to hold that position, it is not hard to imagine a situation in which players or clubs don't trust the ultimate arbiter of appeals, calling into question the whole bureaucratic structure.

In the end, a "good enough" process may be just that. After all, very rarely does any one hit, error, or passed ball make that much of a difference. The magic trick is that, when aggregating statistics, most people manage to ignore the fact that each hit was at first simply a judgment call based on one person's observation of the highly choreographed movement of bodies through space. Neither rules, nor coordination, nor training, nor recourse to an appeals process can completely overcome the fact that in some cases even the best scorers' judgments will not agree. Through the immense work of creating a system—hierarchical, bureaucratic, and disciplined—scorers' judgments have come to be treated as objective and stable facts about the world.

4

From Project Scoresheet to Big Data

In 2014, Major League Baseball introduced a new source of data. The movement of every player and ball would be tracked within inches. Reaction time, angle of inclination, exit speed, and efficiency of path would become the new tools of analytical wizards. It was no longer enough to pretend that assists were a measure of a shortstop's agility, or that slugging percentage somehow captured the essence of power. Now, MLB's Statcast promised data on a level never before imagined. Baseball statistics had finally become "big data."

There is considerable debate in many fields about whether the arrival of big data heralds a fundamental shift or represents a gradual improvement on past computational techniques. Though big data is notoriously difficult to define—beyond the unhelpful "study of large sets of data"—the emergence of data science and data analytics as disciplines, combined with the undeniably rapid increase in both storage capacity and processing speed, has led observers to claim we're in a qualitatively different era of data. Indeed, some baseball analysts, like Keith Law, saw

Statcast as heralding a revolution, a fundamentally new way of understanding the game.[1] As with many instances of big data, there was undeniably something new and different about Statcast, in particular its ability to answer new kinds of questions that weren't even conceivable before. But the actual tools and practices didn't change overnight. The technologies of radar and visual tracking on which Statcast relies were not deployed from scratch but added onto an existing data capture operation. To create Statcast, people had to be trained and organized, tools had to be developed and honed, and calculations had to be verified and stored—and to meet these challenges, MLB relied upon existing infrastructure and expertise.

Nevertheless, there was a long road between official scorers' tracking of hits and errors in 1980 and the emergence of Statcast three decades later. During this period, there was increasing discontent among a faction of fans who felt that baseball data simply weren't capturing the right things. Official scoring had been more or less fixed as a practice, and Pete Palmer was preparing to publish his computerized database of historical statistics. It was known with some confidence how many doubles Nap Lajoie had hit, and even on what days they had been hit. But if someone wanted to know what happened on those doubles—whether they were hit off right-handed pitchers or in runner-advancing situations—the data weren't there. So any questions that might involve strategy were effectively unanswerable.

The measurement of how individual plays affected the likelihood of runs and wins formed the basis of nearly all new baseball analytics in the 1970s and 1980s. Was a walk really as good as a hit? When should teams give up a run for a chance at a double play? Was a high-average hitter more valuable than a power hitter? To have enough data to analyze strategy, however, someone had to collect not only enough examples of similar situations from the deep past to get an adequate sample size, but also enough examples from the recent past to get a sense of the present conditions. These were interrelated projects as far as the technology of data collection was concerned. Whether dealing with a game in progress or one that was decades old, data collectors had to figure out how to translate the game into a series of text-based symbols, compile them, and turn them into something that was amenable to probabilistic calculations.

The first public effort to gather play-by-play data was led by Bill James, as a result of a plea for help that he made in his 1984 *Baseball Abstract*. James requested volunteers help keep play-by-play accounts of baseball games, accounts that could then be aggregated and made public. Such an effort, he promised, could make enough data available that someone could measure the relative importance of singles and walks, provide accurate ratings of players' contributions, and quantify players' overall effect on games won. Volunteers flooded his mailbox with offers, and Project Scoresheet was born.[2]

Though James was the face of this effort, and most accounts of the rise of baseball analytics—Michael Lewis's *Moneyball* and Alan Schwarz's *Numbers Game* still the best among them—make him its protagonist, doing so obscures an important link. One reason an increasing number of people were asking questions about the level of detail in baseball data after midcentury was the success of the various mathematical sciences of modeling and prediction. In the development of these sciences James was the exception, the graduate school dropout and factory night watchman who became the father of baseball analytics. Nearly everyone else involved in this effort had scientific or technical training. George Lindsey, one of the earliest researchers into play-by-play questions in the 1950s, tellingly published his largely ignored findings in the technical journal *Operations Research*.[3] Linear programming, game theory, probability analysis, information theory, cybernetics—these sciences decisively demonstrated that making predictions was a matter of gathering the right sort of data and using the right analytical tools. Before the 1980s, however, data on the effects of individual plays in baseball were not widely available or were segmented into the work of isolated researchers.

Within a year of its creation, Project Scoresheet volunteers were spread out at stadiums across the country, and the group's first publication covered the 1984 season. Now, for a small fee, anyone could have access to play-by-play data. Less than a decade later, however, the project was dead, even as real-time baseball statistics became ever more widespread. Project Scoresheet's rise and fall is usually read as a cautionary tale: once it was discovered how much money could be made selling real-time baseball data, a volunteer project couldn't succeed. Those running Project Scoresheet would abandon it and its nonprofit ethos in order to expand STATS, a private for-profit company once run alongside Project Scoresheet. Eventually, STATS would become very profitable selling baseball data, while Project Scoresheet was left to wither and die.

There's truth to this version of the story, but treating Project Scoresheet as quixotic or naïve doesn't capture the effort's true significance. In fact, parts of the project remain hardwired into Statcast and other twenty-first-century data initiatives. Its historical significance lies in its infrastructure, the methods of organizing people and packets, humans and machines, that have made possible the rise of modern baseball data.

———

Project Scoresheet required more than just volunteer scorers in the stands. After all, the usual ways of keeping score don't capture play-by-play action. Each scorecard entry typically records a player's time at-bat and possible progression around the bases. At the right side of the rows is usually the summation of each person's offensive statistics, and at the bottom of each column the sum of the team's offensive statistics for the inning. Each can be summed again to give the total performance. (The defensive statistics are generally included here as well, of course, just as inverses: hits become hits allowed, ground ball outs become put-outs, etc.)

Going back to the earliest days of scoring, this practice was ultimately about recording individual merits and faults. The

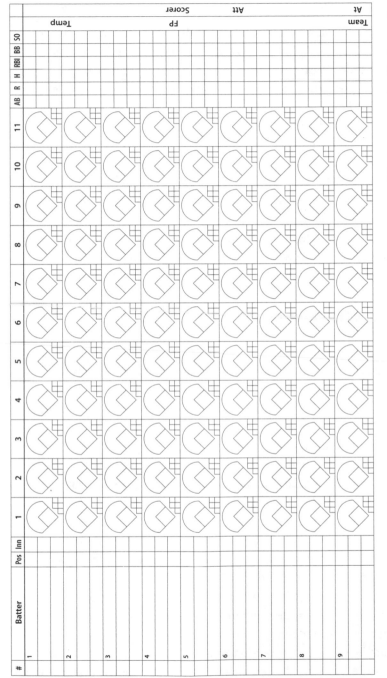

FIGURE 4.1: Traditional scoresheets emphasize individual performance over play-by-play accounting by using each box to represent the offensive contribution of a player. Designed by Christopher Swingley (2005), Creative Commons License, http://swingleydev.com/baseball/scorecards.php

traditional scoresheet didn't provide a convenient way to note the sequence of plays, however. If a leadoff batter hit a single and then stole a base later in the inning, keeping track of that steal in his original row (as "his" steal) wouldn't make it clear what the pitch count was when that happened, or who was at-bat, or how many outs there were. The focus on individual performance would obscure the chain of action and consequence.

By the 1980s, James and others had determined that in order to measure how plays and players influence outcomes, a sequential account was needed. Craig Wright had designed a scoresheet back in the 1970s for just this purpose. Wright, who would go on to earn the distinction of being the first labeled "sabermetrician" employed by the major leagues, wanted a scoresheet to actually reflect the order of play in the game. His insight was to rearrange the sheet to encode activity around plate appearances. Everything that happens in baseball happens with someone at bat—though of course the event might happen before a pitch is thrown or after a ball is hit. So instead of organizing the sheet around individual offensive performances, Wright organized it sequentially around at-bats. A stolen base was marked not in a box tied to the player involved, but in a box tied to the plate appearance when it happened. The order of events and the relevant context for them were in this way much more easily revealed.

Wright numbered every plate appearance, with three horizontal sections for each: the top for things that happened before the main event of the appearance (stolen base, passed ball), the middle for the main event (walk, hit, strikeout), and the bottom for things that happened after the main event (runner moved from second to third base). Pitching changes and defensive substitutions were linked to the exact plate appearance (e.g., Jones replaced Smith as pitcher during plate appearance #21). One telling detail of this new sheet was that the columns weren't organized around innings, but rather around times through the order. A player's row revealed what he did at bat but didn't usually include situations in which he stole a base or scored a run. That's not to say that someone

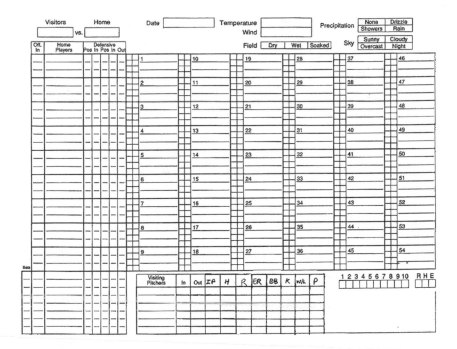

FIGURE 4.2: Project Scoresheet volunteers used a form with three rows for action before, during, and after (numbered) plate appearances, creating a sequence of inputs that could be translated into computer code. Courtesy of David Smith and Retrosheet

couldn't reconstruct individual performance statistics, but that was no longer the goal of the sheet. The point was to track how one play affected the status quo ante, event by event.

Wright showed his sheet to James when they happened to be sitting together in a spring training game just before Project Scoresheet got underway. Because it foregrounded the order of events and the interconnection of causes and effects, it would provide the basic template for Project Scoresheet's play-by-play analysis.

Though Wright did not design his scoresheet for use on personal computers, it was a format well suited for the input of computerized data: instead of trying to put in all the detail around a particular person's overall contributions (he was at bat twice, made one out and one single, with a stolen base), it transformed the baseball game into a series of inputs. All that was needed to encode a game on a computer was input sequences—once, that is, James could be convinced that Project Scoresheet should rely on computers.

Initially, Bill James had hired Kenneth Miller to manage Project Scoresheet, and one of Miller's innovations had been to push for computerization of these data. Toward that end Miller bought four IBM PCjrs in 1984, the year they were released, so that multiple people could input games into memory, one person for each of baseball's four divisions. Though Project Scoresheet didn't need to use computers—volunteers still recorded games by hand—Miller and others involved recognized that with thousands of games, computerized inputting would facilitate year-end aggregation.

Nevertheless, a lot of work had to go into making a scoresheet suitable for computerized input. David Nichols, Geoff Beckman, and especially Gary Gillette worked to formalize precisely how to fill out the scoresheet so that the inputting would go seamlessly, deciding on a system of notation that the scorers could use that would be easy for inputters to follow. Otherwise, each inputter would have to decipher each scorer's notation and style. To create computerized data, Project Scoresheet needed to standardize inputting syntax.

Over time, this group enabled the system to record pitch types, as well as the trajectory of batted balls. It was then possible to indicate not just that at some point in the fifth inning Jones got a double, but that he doubled on a hit to deep left field just after White had stolen second, on the fourth pitch of the at-bat, with the count 2-1, and that White scored from second as a result. Each event became a discrete input.

Once these data were in the computer, it was much easier to print out play-by-play information for the game. After all, the game was now just a series of plate appearances and events. When Project Scoresheet published its record of games, it did so with "account form box scores," which provided a way to see not just what happened, but the game situation in which it happened. For example, say that on July 17, 1984, the catcher Bruce Benedict, batting second, came to bat three times: leading off the third inning, when he grounded out, shortstop to first base, to make the first out; then in the fifth, with one out and no one on base, when he flied out to the right fielder; and then leading off the eighth, when he grounded out to the third baseman. His line would read:

2 Benedict 3a.63 5b.9 8a.53

This string of information doesn't list everything, but it does encapsulate a lot more than the old version of the box score, because it puts the hit, out, or error in context. Someone could even figure out the pitcher during each at-bat, because the pitcher's line would also indicate the number of plate appearances he was on the mound.

By 1985, John Dewan had been named executive director of Project Scoresheet. As an actuary, Dewan had gone to school in math and computer science and was trained to be detail oriented. The collection of baseball data was to him not that different than the practice of an actuary: it involved the analysis of patterns. After a fellow actuary introduced him to James's *Abstracts*, he was immediately hooked and wanted to volunteer his services. As one did at the time, he called directory assistance in Lawrence, Kansas, and reached James's research assistant, Jim Baker, who put him in touch with Miller. When Miller left Project Scoresheet, Dewan took over the software for inputting games—he had been the first guy on his block with an Apple II computer—and with his wife, Sue, who was also trained in computer science, ran Project Scoresheet out of a spare bedroom in their house from 1985 to 1987. Though they both kept their day jobs at the start, the couple soon became more involved with their work on another baseball data project called STATS, in which the Dewans and James had invested in 1985.

STATS had been founded on the back of a different system for entering games into a computer in real time, that of Dick Cramer.

With a day job as a mathematical chemist involved in drug development, Cramer had long been interested in modeling and baseball, and when the opportunity arose in 1981 to work on a system for baseball computer inputting, he got on board. Working with only a couple teams in the early to mid-1980s, Cramer developed a system that enabled a surprising level of detail to be recorded, including the location and type of batted ball, checked swings, and every usual statistic. But by 1985, the system was floundering for lack of interest from teams, and Cramer needed a new infusion of capital, which the Dewans and James provided. Since STATS and Project Scoresheet were essentially doing similar things—taking play-by-play data and storing them electronically—their connection wasn't a stretch. Nevertheless, it is fairly remarkable that to some extent, STATS and Project Scoresheet, two of the most important collectors of play-by-play data in the mid-1980s, were at one point both run out of the Dewans' house in suburban Chicago.

To focus entirely on the few people at the top, though, misportrays Project Scoresheet's organization. After all, to record major league games means having a volunteer observe every game. This isn't a minor task. There were 26 major league teams scattered across the country when the project began, each of which had 81 home games. Team captains were assigned to each local group of scorers to ensure that a volunteer was available and able to cover every game. Ideally, there would be two scorers to do each game, in case one missed a play or mischaracterized a pitch. This two-person coverage was often facilitated by having one scorer for each team—the home-team scorer, ideally—working from the stadium while the away-team scorer worked off the radio or television. Project Scoresheet's 1984 book credits nearly 40 captains, division captains, and coordinators alone, and though not officially tallied, there were well over 100 volunteers in total.[4] Nevertheless, there were still problems with coverage, and cities with an established team like Detroit might have had a dozen volunteers scoring games, whereas captains for cities like Montreal would have trouble finding more than a couple people. Teams like Atlanta, because of the national cable television broadcast by the Turner Broadcasting System (TBS), could sometimes be covered by a volunteer from a different city, though these volunteers were rarely as motivated to carefully record every game if they didn't care about the outcome.

When Project Scoresheet volunteers disagreed, it was not a simple matter to adjudicate. Even if games had been broadcast over radio or television, that didn't mean that a recording existed or, if one did, that volunteers had access to it. As a result, team captains would use the published box scores to try to double-check volunteer scorers' accounts. When official scorers changed their ruling after the fact, volunteers would also have to manually correct them. If all else failed, Gillette remembered, a captain or director of the project would have to beg the team's Public Relations Department or rustle up a scoresheet from a reporter to try to reconstruct the situation. Accurate data here meant not just ensuring multiple people were watching the game, but also creating procedures for reconciling differences between those accounts.

There had always been problems with getting every game covered, but by 1987 Project Scoresheet was becoming increasingly untenable. Volunteers were not thrilled that STATS seemed to be cashing in on their labor. Though the data might support two operations—it was certainly possible to have scorers record the games for STATS to sell while sending a duplicate to Project Scoresheet—that didn't resolve the tension of having the Dewans and James also running STATS: Was Project Scoresheet fundamentally a volunteer, nonprofit public good, or not?

During and after the 1987 season, the tensions made further cooperation unsustainable, and James and the Dewans left Project Scoresheet to focus full-time on STATS. Gillette took over the operation for the 1988 and 1989 seasons, and was faced with figuring out how to collect the pieces and get things moving again. With some volunteers still committed to the nonprofit nature of the endeavor, Gillette reconstructed missing games' data, organized the remaining volunteers, and kept the project going until it folded from a lack of support after the 1990 season.

Project Scoresheet's rapid rise and fall might be taken as a sign of naïveté: data collection was too lucrative to be left to volunteers.[5] Certainly, from the perspective of James and Dewan, the project was a kind of bridge, enabling them to combine the real-time computerized data entry system of STATS from the early 1980s with the ambition of a national network of scorers able to track every play in baseball, day by day. The data they produced emerged out of their ability to construct successful infrastructures for creating reliable data.

———

Still, there are, I think, better ways to understand the lessons of Project Scoresheet in the long history of baseball data, and in particular its influence on subsequent data collection efforts. One legacy was the way Project Scoresheet showed that national data collection efforts required systematic coordination of people and technologies. There was nothing inevitable about the link between computers and play-by-play data. James didn't have a vision of computerization when he put out the ad to start the project. Nevertheless, just about everyone *but* Bill James who was deeply involved with Project Scoresheet and STATS—Dick Cramer, John and Sue Dewan, Gary Gillette, Ken Miller, David and Sherri Nichols, and Dave Smith—possessed computing skills. No doubt, James's full-time commitment to baseball data in the early 1980s was a luxury. He had made some money on his publications and could afford a full-time research assistant, so he didn't initially see the benefit to computerization. Everyone else initially had day jobs, often technical jobs that enabled them to recognize the possibilities that computers offered for data analysis.

James's 1984 call for free and open data for researchers was less compelling by 1990. That six-year span marked a major transition in the availability of up-to-date baseball statistics. James's first *Historical Abstract* was published in 1985, Elias Sports Bureau responded with *The 1985 Elias Baseball Analyst*, and Pete Palmer and John Thorn published *The Hidden Game* in 1984 and *Total Baseball* in 1989.[6] None of these works focused on play-by-play data per se, but they certainly were evidence that baseball data collection could be monetized.

When the Dewans and James invested in STATS in 1985, the money went to equipment, specifically a Digital Equipment Corporation (DEC) MicroVAX, which had just been introduced commercially. At this time the possibilities of networked computing were only accessible with a machine that could do "packet switching" directly, enabling communication from various ballparks to be transmitted to one mainframe computer. When Cramer had first written STATS' entry program, he had assumed that every data inputter would be schlepping an Apple II up to the press box of clubs that were paying for the data—which, after all, were going to remain with the teams. Now, with the appearance of DEC's peer-to-peer network, Cramer could connect scorers, via modem, to a mainframe computer. STATS wasn't succeeding as a service selling teams their own data—the data needed to be mobilized for the company to be truly valuable.

The year after separating from Project Scoresheet, STATS scaled up to provide statistics overnight so that each morning its database had up-to-date statistics from every game. As Schwarz's *Numbers Game* showed, the change, seemingly one of degree, was actually one of kind. Overnight updates enabled STATS to be useful to daily newspaper and wire services. Before the 1990 season, STATS got its first major contract: *USA Today* wanted it to provide daily box scores for its national paper. STATS promised not just accuracy, but also ease. Previously *USA Today* had kept a stringer in the press box who would phone into the office and read off the box score, number by number, which would then have to be transcribed and disseminated. Now, STATS could send a score at the end of the game with a punch of a button, simultaneously updating the season totals in the database. Within two years, STATS also had contracts with ESPN and the Associated Press, and it temporarily surpassed Elias as the premier provider of statistics for media outlets, even if it did not reach the level of providing Major League Baseball its own records.

Dewan hired enough scorers to ensure that at least one would be present at every game, in case television or radio broadcasting went down. Each reporter (STATS referred to them as "reporters" rather than stringers or scorers) had to undergo extensive training and then a stint as a backup, or apprentice, before being promoted to the main reporter in the stadium. Each STATS reporter also had to score on paper before entering anything into the computer. As soon as a foul ball was hit, for example, the reporter in the stadium and the backup elsewhere would both immediately mark "F" on their paper scoresheets, and only then would they punch in "F" on their keyboards. Seconds after any event, there would be four records of it: two within separate computer memories, and two on the paper scoresheets.

After the game, the records would be sent to Dewan at headquarters, and another program would run through the two computer entry sequences to see if there were any discrepancies. The original software Cramer designed blocked logically impossible entries, but it was possible to have reporters disagree on questions such as whether a particular out was on the ground or in the air or if the first pitch had been a ball or a strike. Any discrepancies would be flagged, and then another staff person would have to research them, perhaps by calling the reporters, looking up newspaper accounts of the game, or asking club officials. Eventually STATS introduced a third account so that a discrepancy could be more quickly resolved simply by taking the majority ruling of the three reporters. These problems were very similar to the ones that Project Scoresheet had faced, and solutions to them again required a substantial infrastructure to be built.

Such error-checking mechanisms were useless if the data never arrived at headquarters. Modems in 1990 were hardly reliable. Reporters were warned by the STATS' manual that there were "many hardware elements in successful communication, and a press box scoring session involves five separate phone calls." After all, "computer communications can be a very sensitive procedure." Little hints were provided, like the fact that the Toshiba T1000 modem would be hard to reset after being turned off, so reporters should first try to unplug the phone line for a few minutes and then retry.[7] By 1990 these communications were happening each half-inning—hundreds of them every day, coming from all around the country. Every operator had to be familiar with emergency numbers.

Both the technology and the reporters—the packets and the people—needed constant massaging, monitoring, and tweaking. Project Scoresheet provided a blueprint for the infrastructure of STATS' national scoring efforts, a way of linking human data entry processes to centrally stored electronic records. Only with this infrastructure in place might the technological advances be leveraged to produce the real-time, national, verifiable data that would characterize the growth of STATS by the early 1990s.

A second legacy of Project Scoresheet was the epiphany it planted in the mind of one of those closely involved in the organization's final year. Project Scoresheet and STATS were effectively collecting play-by-play data for every current game, but for the vast majority of games that had ever been played, no one had any record of what happened play by play, let alone pitch by pitch. Dave Smith began to wonder about this gap, about all the games played before Project Scoresheet was started. Official box scores didn't exist for those games, only the league-required statistical summaries and daily stories in the newspapers. But if the question was whether Nap Lajoie hit a double in the second inning or the fifth on a day he went one for four—not to mention who was pitching or whether there were runners on base—it was not easy to figure out. Smith had the idea of solving this problem by collecting the play-by-play data for every game played since 1871—a giant undertaking but, in theory, a doable one. He called it Retrosheet.

When he left Project Scoresheet after 1990, Smith was still working as a biologist at the University of Delaware. He had gotten into biology because it enabled him to do the sorts of things professionally that he wanted to do with baseball on the side; it wasn't that his day job gave him skills for his hobby, but that both used the same skills and one happened to pay the bills. Project Scoresheet had collected the data from games starting in 1984, so Smith took 1983 as Retrosheet's last year and 1871 as its first. Smith's idea was to work backward from 1983 and figure out how each game had unfolded, one play at a time. Baltimore wasn't far from his home in Newark, so he started by contacting Eddie Epstein, an employee of the Orioles who was sympathetic to new baseball data initiatives. Epstein helped Smith get the team's scorebooks from 1954 to 1983, and Smith was off.

Twentieth-century baseball clubs often collected scorebooks in order to have a record of their own performance for publicity or historical reasons. There was no requirement to do so, and not every team kept them or saved them when changing cities or facilities. But when team scorebooks existed, they were the first source of data for Smith's new project because they provided play-by-play information that neither box scores nor statistical summaries could give. Smith went from team to team, trying to convince someone, anyone, in public relations to dig out the old scorebooks. Often, he found that employees lower down in the hierarchy would be more willing to let him copy the files; PR directors might be suspicious of his motives. On at least one occasion, he waited until the team was out of town and then visited the offices when only a skeleton crew was present in order to convince a friendly person to allow him access. Over time, PR people started vouching for Smith and Retrosheet, and in less than a decade he got access to whatever historical scorebooks existed from every club.

After examining clubs' scorebooks, Smith and a group of volunteers turned to the scorebooks of reporters, announcers, and fans—pretty much anyone who might have been at games—as well as old newspaper play-by-play accounts and archived broadcasts. Sometimes the people themselves didn't realize why they had kept the scorebooks. One story Smith remembered fondly involved Bob Stevens, a retired sportswriter for the *San Francisco*

Chronicle. After digging out 30 years' worth of scorebooks, Stevens told Smith, "All these years I never knew why I saved my score-books. I guess I was saving them for you."[8] Often Smith was not so lucky, and teams or writers had already trashed their books: the Braves tossed out nine years of scorebooks after they moved to Atlanta, the Red Sox after a renovation of Fenway Park in 1968, and the Dodgers when they moved to California. The St. Louis Browns apparently saved nothing at all after 1953, when they moved to Baltimore. Over time, sources have become harder and harder to locate, though occasionally a new scorebook appears on eBay—a volunteer scours the site daily for Smith—or is donated to the Baseball Hall of Fame after being discovered in an attic. Today Smith is well known enough in baseball circles that he'll likely hear if any new sources are found.

It wasn't the case that the older games were harder to track down. In reality, source availability depended more on historical contingencies than on proximity to the present. In certain decades, cities had multiple competing daily papers, each of which would provide a game account that might give details of scoring plays or—ideally—describe the afternoon game play by play for evening commuters to read on the train. As newspapers disappeared and night games made evening newspaper accounts irrelevant, these records became thinner, not least because the number of reporters keeping score diminished in turn. The emergence of neither radio nor television fully replaced these sources, for only in exceptional cases do audio or visual recordings of games exist.

Even if Smith found two newspaper accounts, they might have box scores that differed slightly in their particulars, neither of which might match the scoresheet produced by a fan at the game. And none of the above necessarily matched the official scorer's account of the game. That's assuming these documents even existed. Retrosheet was a history project masquerading as a data-gathering effort.

Once he had a complete or near-complete game account at hand, Smith used the data entry program that emerged from the embers of Project Scoresheet to record the game play by play, as if it were happening in real time. Of course, he rarely had pitch-by-pitch information, or even information on the count prior to a batted ball, but when he did, he recorded it. Once the game was entered, it went into a database, where it was possible to figure out players' and teams' records and compare them with the official records. This required extensive proofing of each year of games—something on the order of 500 hours of work *after* every game had been entered—and comparison with the day-by-day records put out by the leagues. Then Smith, in the tradition of Project Scoresheet, released the information for free at retrosheet.org. Anyone could—and can—download the entire dataset or go to some specific day in the past and see what games were played and what happened in them.

Retrosheet volunteers have discovered thousands of what they call "discrepancies" with the official record. This might be because of mistakes in data entry among Retrosheet's volunteers, mistakes in Retrosheet's sources, mistakes in the official scorer's report, mistakes in the way the report was transcribed, or mistakes in the way those transcriptions were aggregated. While sometimes the problem is obvious—a game appearing twice on the official ledger, or data entry that diverges from the source—often it's simply impossible to figure out precisely how the discrepancy emerged. Other than a few researchers who have an obsessive interest in a particular player's record in 1933 or National League runs batted in (RBI) totals in 1912, no one's done this before, so there is little information to go on. As good historians, the project's volunteers acknowledge that it is better to note differences than to simply erase them, but Smith is usually willing to take a guess on why he thinks that the official record differs. One whole section of Retrosheet's website is devoted to these discrepancies.

Discrepancies suggest the ways in which historical baseball data has a materiality very different from that of data in the making. Take the work of one Retrosheet volunteer, David Vincent, who served as the secretary of the organization after the creation of its board of directors in 1994 until his death in 2017. (Vincent was also an official scorer for the Washington Nationals and a long-time official scorer in the minor leagues, so he knew what he was talking about when dealing with baseball statistics.) Years ago, at the Baseball Hall of Fame, Smith came across the scorebooks of Cincinnati reporter Tom Swope, who was integrally involved with the BBWAA over the four decades he covered the Reds. Vincent loved using the Swope records, because they were "precise," "accurate," and "just a joy to work from." Other sources' scorebooks were a "horror story," and the difference was significant: instead of taking three hours to decipher a fan's scoresheet and record the game, it took only about 15 minutes to enter in Swope's data and another 5 to double check it.[9]

Data are not free floating, but tied to the material record. During a Reds game in the 1940s, Swope might have kept score on paper for the *Cincinnati Post* while the official scorer filled in the written report for the National League. (Sometimes, presumably, Swope would have played both these roles.) After the game, the official scorer's summary would have gone to the league and been written into the day-by-day ledger. Seventy years later, Smith went to the Hall of Fame, took a digital picture of Swope's original scorebook, and sent the file to Vincent, who then used Retrosheet's software to enter the game into the database. Because the day-by-day ledgers have been microfilmed and entered into a computerized database, Swope's scores can finally be compared to the official scorers' summary (something that would not necessarily have ever been done before). The statistics are not that different than dinosaur fossils or other specimens used in data-driven sciences—they have to be carefully collected, cleaned, compared with standards, fit into patterns, and documented.[10]

Retrosheet volunteers must also deal with situations in which the data don't exist and aren't likely to in the future. Tom Ruane, a longtime employee of IBM who stumbled on Retrosheet in the late 1990s through his involvement with SABR, came up with one solution to this problem.[11] Ruane's focus has been on games without play-by-play data but for which it might be realistic to get a box score. Obviously, for the games for which Retrosheet had the play-by-play data, creating the box score was a trivial matter (indeed, the entry program could compute it automatically). But for games for which the play-by-play data may have never existed, Ruane still wanted to have the box score. He thus created a new kind of file with all the details of who played and the record of their performance. Starting in 2003, he arranged the posting of box scores for a few 1963 Kansas City home games and over the next decade completed season after season, until now there are box scores for every game going back more than a century, regardless of whether the play-by-play data for them have been located.

For games about which more data existed, but still not enough to give a full account, Retrosheet designed a new "deduced" game file. A box score provides a lot of information, but often it isn't clear when hits or outs were made or when stolen bases occurred, especially if they didn't contribute to the scoring. Tom Thress has taken on the job of deciphering the action of these games, deducing about one game a day. He got into this work because it is effectively a logic puzzle without a unique solution—he's always been "fascinated by history," and this offered a way to figure out precisely what might have happened in a game when the evidence is missing.[12] For a given game, Smith sends him a set of newspaper articles and the box score. Thress sets up a spreadsheet that sets out the logical parameters of the play-by-play. If a player has five plate appearances and five doubles according to the box score, his contribution is easy to figure out: every time he came to bat, he hit a double. Most appearances are more complicated to decipher, but it is usually possible to narrow each down to a few possibilities, which can be narrowed further when combined with the pitchers' and defensive records. Ideally, what's left is a play-by-play account with only a few missing details.

Again, the skills Thress has brought to bear on this task are grounded in those he uses in his work. Models of what might have happened are what Thress is good at—his day job involves modeling demand for the U.S. Post Office's volume and revenue. There's a set of parameters and requirements, from which one must try to guess what's really going on. Of course, his day job looks forward and his hobby to the past, but they are both concerned with trying to fit the data to a model.

Creating Retrosheet's data required managing sources, conventions, and technologies of data entry, as well as initiating forms of quality control. But this data problem was intertwined with a people problem. Retrosheet remains an all-volunteer effort, and hundreds of people have contributed over the years. Nevertheless, the bulk of the effort has really been done by a handful of individuals: Smith, Vincent, Ruane, Thress, and a few others. (Another early contributor, Sherri Nichols, also played a central role in Project Scoresheet, alongside her husband, David.[13]) Such a small group has its advantages. It is easier to reach agreement, and volunteers can take more ownership over their own products. Likewise, Smith is more or less a benign dictator. He's charismatic, gregarious, and almost universally well liked. Such qualities were essential to convincing teams and writers to cough up sources early on, when the project was nothing more than a vision, and it remains essential for herding volunteers.

Retrosheet now features play-by-play data for games going back over eight decades, box scores for games going back more than a century, and some information on nearly every professional game, player, manager, and umpire since 1871. Through various partners, the site also has information from 1984 on, though this was technically not the "retro" part of the effort. Today, the website is useful but utilitarian. It is a source for delivering the underlying data, not for quickly looking up a statistic. To do the latter, everyone heads to baseball-reference.com, but don't be fooled by the latter's flashier interface—what you'll often find there are just records drawn from Retrosheet and Pete Palmer's database.

If Smith hadn't spearheaded this effort and carefully developed it over the last 25 years, or if he had been less successful at cajoling others into joining his effort, it is hard to imagine anything equivalent succeeding. In a sense, Retrosheet has quietly changed the ability of people to do historical baseball research. Many analytical questions that rely on play-by-play data from more than the recent past wouldn't be answerable without Retrosheet. The flashy new metrics may get the attention, but researchers know they depend on Retrosheet's reliability and thoroughness to design and test new tools. All Smith asks is that researchers include a brief note in papers that use his data indicating that the information came from Retrosheet. More than one grateful researcher has added to this, "God Bless Retrosheet."[14]

On its surface, Retrosheet appears quite different from Project Scoresheet. The former is concerned with the past, the latter with the present. Smith himself identifies the origins of Retrosheet as not the late 1980s, but his first Dodgers game in 1958. It is, in a sense, the perfect encapsulation of baseball nostalgia in the age of data analytics: the very same effort that aims to turn every baseball event in the twentieth century into downloadable data has also enabled fans to relive the action of any game they've ever attended.

Nevertheless, there are clearly also similarities between the two. Project Scoresheet and Retrosheet have both been obsessed with the problem of creating reliable data from the available evidence. They have required the organization of people to gather evidence and technologies to record and compile that evidence. Discrepancies have had to be dealt with, as have missing or incomplete data. Project Scoresheet didn't really fold when STATS took over the bulk of its data-gathering efforts. Some of its members simply turned to a different challenge, one that would likely never attract much commercial interest: the recovery of the data of baseball's past.

————

The third legacy of Project Scoresheet has been more invisible, if more consequential. The code and data entry system built by the scrappy Project Scoresheet crew—who initially had to beg team officials just to help them fill in missing data—would eventually provide some of the infrastructure for Major League Baseball's own data operation less than 15 years later.

When the Dewans split from Project Scoresheet to devote their attention to STATS full-time, they took with them Project Scoresheet's data entry software. Without this software, the project consisted of a bunch of filled-in scoresheets gathering dust. Luckily, Tom Tippett happened to be active in Project Scoresheet at just this time.

Tippett, who grew up in Toronto's eastern suburbs, was continually on the cutting edge of computing. While in high school in the mid-1970s, the school district bought one computer for all 13 high schools, and he joined a new programming club. In order to run a program, the club's members would have to create the punch cards and then give them to the teacher each week to process on the central computer and bring back the compilation errors and other results to be fixed the next week. This was not high-speed computing. At the University of Waterloo, he was formally taught programming and gained access to a leading academic data center. Significantly, he arrived there just as punch cards were being phased out in favor of entry terminals and started using an IBM Series/1 computer to compile his programs, dropping his turnaround time to minutes, rather than days. After his 1982 graduation, with a job at IBM in downtown Toronto, he not only discovered Bill James but also had the idea to create a computer-based baseball simulation game.[15]

Tippett developed a prototype that year, but it took over three years and an MBA from Harvard for him to begin turning his prototype into a commercial product. The end result, which came to be called Diamond Mind Baseball, needed to be able to work even if people didn't have much computing power, so he designed the program to fit onto one 5¼-inch floppy disk, with a second floppy disk to hold player data. Users would fire up the program on the first disk, insert the second to choose teams and players and load the data, switch back to the first to play the game, and then switch back to the second to save the stats from the game. The core of the simulation was a relational database that stored the probabilities of all possible events in the game based on past statistical frequencies.

It was a happy coincidence for Tippett that while he was designing his game, he discovered Project Scoresheet. After all, his game was going to be based around detailed situational data, and that was Project Scoresheet's specialty. Previous baseball simulation games, like Strat-O-Matic, had used dice or spinners to randomize outcomes and then used a series of cards to connect those outcomes to baseball events based loosely on a probabilistic model of the game. Tippett wanted his system to be far more detailed—able, for example, to take advantage of how good various fielders actually were, to distinguish between left-handed and right-handed batters' situational outcomes, and to take into account where the game was being played. Working with this level of detail using cards would be infuriating, requiring lots of cards and even more time to look up every detail. Electronic computers, however, could sort through this information with extraordinary speed: the simulation could be made immensely complex and yet the outcomes still calculated fast enough and easily enough to finish the game in about 30 minutes. Strat-O-Matic had proved that a computer wasn't needed to model baseball, but with one, Tippett realized, there was the possibility of much finer-grained detail.

Project Scoresheet provided Tippett the data he needed. Within a year or two, he had access to every game in Project Scoresheet's database, and in computerized format, which meant he could use these data to program in probabilities for his own game's models. When Project Scoresheet threatened to come apart at the seams in 1987, Tippett had the incentive to keep it going.

He also had the know-how to replace the original input program for games, and he had a partner in David Nichols, then a doctoral student in computer science at Carnegie Mellon, as well as Smith himself. Tippett took all of his vacation in the spring of 1988 to design the interface for the software while Nichols figured out how to parse the inputs and generate the game data. They designed the basic inputting system for Project Scoresheet in just over two weeks.

They had two advantages in these efforts. First, Tippett had already designed an interface for Diamond Mind Baseball, and since he owned that entirely, he could borrow freely from it. Second, the team agreed to keep Project Scoresheet's entry syntax, which had been formalized years earlier using Craig Wright's design for scoresheets and its later adaptation for coding. Gillette and others had taken Wright's basic system and expanded it to include not just more detail (pitch selection, catchers' interference calls, pitchouts, and so on) but also the particularly rare or complex plays that still had to be coded. This work enabled Project Scoresheet to turn the messy, chaotic, and often unpredictable events of a baseball game into a standardized string of letters, numbers, dots, and dashes.

The syntax, though a bit cumbersome at first glance, had the advantage of economy—no mean advantage when navigating the problems of storing data on dozens of teams, hundreds of players, thousands of games, and hundreds of thousands of individual events in the late 1980s. Each defensive position received a number 1 through 9, and each offensive event a letter. A fly out caught by the center fielder could be entered solely as 8, and a ground out at first on a ball hit to the shortstop as 63. "Modifiers" were indicated with slashes, while a dot indicated advancement of runners and a semicolon simultaneous action. If the play was in fact a fly to center caught by the center fielder deep in center field, it could be entered as 8/F8D. A sacrifice fly to deep left field with the runner on third tagging up and advancing to home would be 7/SF/F7D.3-H. Parentheses could specify who was retired on the out, as well as further details of the play: E5/G.3XH(5E2);B-2 meant a third baseman (5) booted (E) a ground ball (/G) and then threw home trying to get the player running from third to home (.3XH); the catcher then dropped it for an error (5E2), but it would have been an out, meaning that the third baseman should get an assist, and during that

Hit Scoring System Notation

HIT CODES		THROUGH-THE-INFIELD HIT EXAMPLES	
Single	S	Down the 1B line	3L (D,E3/L3L)
Double	D	Through 1B-2B hole	34 (S9/G34)
Triple	T	Up the middle	46 (S8/L45)
Home Run	H or HR	Through SS-3B hole	56 (S,E7/G56)
		Down the 3B line	5L (D7/G5L)
DESCRIPTION CODES		Directly through or at infielder	1T, 3T, 4T,... (S7/L57)
Bunt	B		
Ground Ball	G	**OUTFIELD HIT EXAMPLES**	
Line Drive	L	Down LF line	7L (D7/F7L)
Fly Ball	F	Short LF (in front of LFer)	7S (S7/L7S)
		Deep LF (behind LFer)	7D (T7/F7D)
INFIELD HIT EXAMPLES		In the LF-CF gap	78 (T7/F78)
Between home and mound	12 (S1/B12)	Short CF (in front of CFer)	8S (S8/L8S)
Inside infield, down 1B line	13 (S3/G13)	Deep CF (behind CFer)	8D (H8/F8D)
Inside infield, down 1B line	15 (S5/B15)	In the RF-CF gap	89 (D,E8/F89)
Past mound, right side	14 (S4,E4(TH)/G14)		
Past mound, left side	16 (S6,F16)		

FIGURE 4.3: Project Scoresheet code allowed varying amounts of detail to be entered for each hit. Courtesy of David Smith and Retrosheet

action the batter-runner made it to second safely (;B-2). Had the throw been late to home, thus erasing the assist, though with the catcher's error still allowing the batter to reach second, the string would change to E5/G.3-H(5E2);B-2. Such code is both specific to action and sequential in time. A whole game could be rendered with about 150 lines of code, taking up about 5 kb of memory—not a trivial amount in 1988, but small enough to work with.

This simple and effective code would have a life much longer than that of the struggling Project Scoresheet. Tippett, Nichols, and Smith created an entity called DiamondWare to hold and license the software. When Smith started Retrosheet, he used the DiamondWare program for computer entry. Tippett and Nichols were by and large done after this point, and Smith became the main person for troubleshooting and debugging, as well as the person responsible for updating the code to be able to handle new inputs over time for Retrosheet.

The past and the present would intersect yet again, however. John Thorn—Pete Palmer's collaborator on *Total Baseball* and *The Hidden Game of Baseball*—joined forces with Michael Gershman in the 1990s to create Total Sports, a company that aimed to expand the publication of statistical databases from baseball into football, basketball, hockey, and other sports. By 1996, Thorn and Gershman not only were publishing print encyclopedias but had moved online. Totalbaseball.com had been named to C/Net's Hall of Fame and *PC/Computing*'s Best 1,000 Websites list in 1996. Their success led to the company's merger with KOZ Sports in 1997.[16]

No one closely involved remembers the subsequent history of Total Sports as anything other than ill-fated. This is in no small part because the company got caught in the bursting of the turn-of-the-century internet bubble. After the merger with KOZ Sports, the new company, based in Raleigh, North Carolina, moved to capitalize, filing an initial public offering (IPO) in November 1999. Less than six months later, in May 2000, as the dot-com bubble crashed, the IPO was withdrawn. In July 2000, San Francisco–based Quokka offered to buy the online component of Total Sports for $130 million in stock.[17] Thorn was able to buy out the print side of the business and started Total Sports Publishing in fall 2000.

Everything failed. Quokka laid off two-thirds of its workforce only a couple months later and was named *Forbes*'s "Disaster of the Day" in February 2001.[18] By April 2001, the company had filed for bankruptcy. Though Thorn's publishing side still owned "official" status with some of the professional sports leagues, it too declared bankruptcy before the end of the year. What had looked to be a lucrative online opportunity in 1997 seemed a colossal mistake by the end of 2001.

Both parts of the story—print and online—are indicative of how data aren't just out there, waiting to be sold. They have to be made useful; their materiality matters. On the print side, Thorn was hired back as an editor by the company that bought the rights to *Total Baseball* to put out a new eighth edition of the book. But the new publisher, Sport Media Publishing, decided to use heavy paper because of the included photographs, turning the book into a 2,700-page, 11-pound beast. Thorn remembered that the joke at the time was that it could be sold with a handle as an "aerobic edition," but you'd have to buy two to stay balanced.[19] The edition was priced at $69.95, and the publisher still lost money on every book. Moreover, Palmer—who had provided the database that underpinned the first edition—had at the time of the eighth edition already agreed to work on a rival encyclopedia, meaning that this edition of *Total Baseball* had to find a new source for its statistics. So, in a supremely ironic twist, Thorn and Sport Media Group had some programmers reformat Sean Lahman's online database, which itself had been reverse-engineered from the CD-ROM included with *Total Baseball* in 1993. The database created by Palmer that had lain at the heart of the first edition of *Total Baseball* had been replaced in the final edition with one reverse-engineered from an unauthorized version posted online.

The online side of the Total Sports saga also stands as a stark reminder of the human labor behind data. Keeping totalbaseball .com updated required having a day-to-day system for capturing baseball statistics, and for this the company turned to Project Scoresheet veteran Gary Gillette. Gillette had continued using the DiamondWare data entry system after 1990 to collect (and eventually sell) real-time baseball data, first for the company Sports-Source and then for his own Baseball Workshop. Despite Gillette's entreaties for Total Sports to use the existing (and well-tested) Project Scoresheet software after the 1997 merger, Total Sports' executives wanted to design their own data collection software for baseball, thinking that it couldn't be that much more complex than the systems KOZ programmers had built for other sports.

Tippett's computer simulation game, Diamond Mind Baseball, had been using data licensed from Gillette to determine its underlying statistical database. After the mergers, Gillette's operation was absorbed by Total Sports, so Tippett began buying the data from them instead. But something was wrong. The new company simply didn't know how to produce baseball data reliably. KOZ Sports' online operation had been founded on datacasting college basketball games, but the software and skills developed for collecting data on basketball were nowhere near adequate for baseball. Data collectors for the Total Sports–KOZ operation were more or less accurate in noting whether a single or double had been hit, for example, but when the system prompted scorers to indicate *where* a ground ball had been hit, scorers would either click on the nearest base or where the nearest fielder normally stood. This would significantly overstate or understate the actual defensive contribution of the fielder. Every ball fielded by a second baseman, for example, would be recorded as an easy play (hit right at him) or as a great play (hit to the second base bag). Either way, the new system provided terrible data. This mattered to Tippett because he included defensive ability as part of his simulation: when the defensive range ratings became wonky, the credibility and realism of his simulation broke down.

Defensive positioning errors were only the beginning. The first year the new Total Sports operation used its own data software, 1998, problems of all kinds quickly accumulated. During the early part of the season, Gillette remembered, Total Sports was sending out a revised software build nearly every day to correct for newly discovered problems. The entry software consistently made simple errors, such as counting the run when a batter grounded into a force out with two outs and the bases loaded. During 1998's exciting home run contest between Sammy Sosa and Mark McGwire, the system couldn't even produce accurate home run and RBI totals for the players. The software was so bad that hundreds of games had to be reentered after the season was over; Gillette hired Palmer and Smith to help clean up the data. Gillette resigned just weeks into the next season, convinced that the company's owners weren't willing to invest the resources required to collect data successfully.[20]

As nearly everyone who has tried can attest, it takes an immense amount of effort to create a reliable infrastructure for capturing baseball play-by-play data. As data collection continued to flounder, Gillette was approached during the 1999 season by a Total Sports engineer who asked him to try to quietly broker an agreement between Total Sports and the owners of the Project Scoresheet software, DiamondWare. (Gillette, in addition to wanting to see the code continue to be used, had a financial incentive to facilitate such an agreement because he still owned stock in Total Sports.) Gillette did so, and Tippett, Nichols, and Smith, who still owned DiamondWare, agreed on a licensing fee and granted Total Sports the rights to transfer or sell the software if Total Sports was purchased. In late 1999, Total Sports started to rely on DiamondWare and Project Scoresheet's tried-and-true data entry program.

What Tippett, Smith, and Nichols didn't know was that after Total Sports' online operation was sold to Quokka the next year, the company—in an effort to cash out assets before declaring bankruptcy—would sell the baseball-scoring software to the newly formed Major League Baseball Advanced Media (MLBAM, or BAM).[21] MLBAM got a steal of a deal: it not only acquired the inputting program, but also received access to Pete Palmer's database, the client-end software Total Sports had developed for disseminating the data, data editors and managers, and Total Sports' extensive network of stringers and data inputters. Project Scoresheet would live on, not as a scrappy operation of outsiders, but as the heart of MLB's own data collection efforts.

———

Major League Baseball Advanced Media has become the acknowledged giant in the field of baseball data. No one else is in the business of data collection to the same extent—MLBAM effectively replaced Elias's data collection operation for major league baseball and now provides the data directly to clubs. MLBAM is also the official statistician of the minor leagues and is essentially the only source for current play-by-play statistics at that level.

MLBAM was formed in the summer of 2000 when baseball club owners voted unanimously to aggregate online rights to baseball. Bob Bowman was hired as the company's CEO later that year, with plans to relaunch mlb.com and live broadcast in-game data with the start of the 2001 season. MLBAM envisioned itself as an independent technology company.[22] The main difference from most tech start-ups, though, was that MLBAM was quickly profitable. Each club committed $1 million a year to the venture for four years, but MLBAM went through less than $80 million before recording a positive cash flow, and since then it has offered the clubs both a handsome dividend each year and a stake in what's now valued as a multibillion-dollar company.

Despite being funded by the corporate barons of baseball, MLBAM acts like a small tech firm, with a relaxed dress code and an open-floor plan at its chic offices in New York City's Chelsea Market. This wasn't an accident. The company did want to attract younger, tech-focused people who wouldn't want to work in Midtown. More importantly, though, MLBAM physically located itself on a major axis of what's now known as Silicon Alley. Ninth Avenue, which runs right outside the building, is a major traffic conduit, but it is perhaps even more importantly a central "fiber highway" of the internet, with Google maintaining offices across the street.[23] By choosing to locate right on this corridor, MLBAM also positioned itself as a major player in internet streaming, and it has handled the streaming logistics for ESPN, HBO Go, World Wrestling Entertainment, and many other services.[24] Play-by-play data may be far easier to distribute with the internet, but the physical infrastructure enabling the movement of data still matters.

Describing MLBAM as a typical twenty-first-century tech firm captures something about the business but gives the incorrect impression that it sprang from thin air in 2000. It is really best thought of as an outgrowth of ventures dating back to the 1980s. The man hired to run the data collection program of MLBAM, Cory Schwartz, had run a short-lived data collection program for the Yankees in the early 1990s and then had worked for STATS as one of its reporters. Nearly two decades later, Schwartz was still in charge of data collection, with the title of vice president of statistics.

Schwartz was not the only holdover from previous scoring efforts. When MLBAM purchased Project Scoresheet's data collection system from Quokka in 2000, it also hired Total Sports' employees to start collecting data. The plan was to have mlb.com go live in March and then start populating the stats section of the website in real time with the season opener. (Prior to the 2001 season, MLB's website was run by Sportsline.com, a media company that also ran the website for PGA Golf and NFL Europe.) That April Fools' Day, the Rangers faced the Blue Jays at Hiram Bithorn Stadium in San Juan, Puerto Rico. For this first test of the system, Schwartz arranged for an old Total Sports hand, Hank Widmer, to run the data system off the television—to be, in MLBAM parlance, a stringer. Widmer didn't miss a pitch, and the action was seamlessly recorded and posted. Mlb.com's first live data feed was flawless.

It turned out to be a fluke. The following day, day 2 of the 2001 season, had a slate of 10 games taking place across the country, with part-time stringers in place for all of them. What unfolded was still, more than a decade later, painful for Schwartz to recall. It was, he conceded, "one of the biggest professional failures I've ever been a part of at any level."[25] Stringers made inevitable entry mistakes, yet there was no effective way to communicate with them in the stadiums or to help them remotely. There was also not a way to monitor what they were entering as they were entering it, so MLBAM couldn't follow what was happening in real time. Nor were there effective tools to make corrections directly from the New York office. And, as MLBAM learned during the Mets' opening-day game against the Braves, when scorers themselves made a correction, the system wasn't clearing out the mistake and replacing it, but instead layering the correction on top of the mistake, causing the statistical records to compound with every press of the "send" button. According to the site, Robin Ventura ended the first game of the season with 92 RBIs.

For MLBAM, the rest of the 2001 season became an object lesson in the ways data don't simply fall out of games but must be carefully coaxed out, cleaned, organized, and stored. Collecting data, MLBAM learned—as had STATS, Project Scoresheet, and Retrosheet before it—posed technological problems. A lot could go wrong. Stringers for MLBAM originally had to use paper scoresheets and then enter their information into the computer, creating redundancy right from the get-go. After all, the internet connection could go down; the laptop hard drive might fail; a power outage could occur. As MLBAM discovered the hard way at U.S. Cellular Field in Chicago, a single foul ball can destroy a laptop. In places where there wasn't a reliable internet connection in the early years, stringers would score on paper and fax the scoresheet to the office, and then Schwartz and his New York team would program it directly into the system play by play in the wee hours of the morning, as if they were the ones watching it, and upload it before dawn as if nothing had happened. By the end of the first season, MLBAM had a stringer in the press box, a backup watching on television elsewhere, and a team of people going through each of the feeds for errors or typos. Everyone communicated using the messaging service AIM in real time, and laptops were equipped with software allowing remote takeover if required. Schwartz's mantra became "capture the data anyway you can"; once the event was over, it was much harder to go back and re-create it.

Data capture—as scoring has come to be known in the twenty-first century—also remained a problem of human labor. As the verb "capture" implies, MLBAM wasn't sitting back and waiting for the data to collect themselves. Widmer didn't reveal the problems with the system on the first day because he made no errors. That was too high a bar—everyone eventually makes an error— but it was clear that collecting data effectively required some degree of baseball expertise. Schwartz eventually gave prospective stringers a quiz asking basic questions about baseball as well as technical questions about the entry system. Prospective stringers would have to explain the "wheel play" as well as determine who was most likely to "cover" second base on a steal attempt with a left-handed batter at the plate. They would also have to know something about official scoring so that they could generally predict the scoring of a situation rather than having to delay entry until the official scorer announced every minor call.

Bodies have to be disciplined and potential accidents anticipated. Stringers can't drink too many sodas or they will risk missing a play while in the bathroom; they have to bring multiple pencils for scoring by hand; they have to bring a cell phone with a fresh battery; they have to have binoculars to see defensive shifts and changes in real time; and above all, they have to have an impressive ability to focus intensely for hours. MLBAM hasn't figured out how to predict these qualities perfectly, so they still send people out for trial games with veteran stringers. Without controlling the body and managing its mental and physical conditions, the data won't be reliable.

MLBAM also, until 2015, ran an eight- to ten-week training program to get every potential stringer comfortable with the entry software that Tippett, Nichols, and Smith had originally designed. Because this software was essentially a language, stringers had to learn to distinguish HR/78/F from 8/F.2-3;1-2. With the 2015 season, MLBAM built a new menu-driven interface, so instead of HR/78/F, users could select "home run," "left center field," and "fly ball" from the menu choices and the entry code would be automatically generated. Schwartz could take someone with the requisite physical and mental abilities and have him or her score a game within a week. This revamp was partly a requirement of the massive increase in the number of games covered by MLBAM since 2005 as it has expanded to the minor leagues as well as the World Baseball Classic, spring training and winter league games, and various independent leagues. The Project Scoresheet language remains the "backbone" of the data-gathering effort, but it can operate in the background.

There's always someone looking over a stringer's shoulder. Schwartz had a full-time staff of about 26 employees (largely in New York) who were dedicated to data collection, cleaning, and production. He likened this operation—only half jokingly—to a casino. There were stringers in the park "dealing" the initial data, five pit bosses (managers) who observed a few games a night to ensure the stringers were doing ok, and the "eye in the sky" (the "adult in charge") who watched the five managers to make sure they were communicating with stringers effectively and helped with any systematic connectivity issues. Then there was the "count room," the final layer of management, a team of people who came in at 4 a.m. to go through the data one more time and do any fine-tooth final edits. By 6 a.m. New York time, the complete data package was ready to be sent out to teams.

While corrections might still be made later—an official scorer's decision might be changed, say—the vast majority of data editing was completed by that point. In addition to his stringers and full-time staff, Schwartz also hired part-timers to come in at the end of the season and go through every single major and minor league game yet again, reading the code page by page and trying to figure out where errors might have been made. Was a hit to the catcher marked as a "hard hit ball"? That seems unlikely. Or did a stringer mix up the order of pitches, which wouldn't affect the official scoring or the play per se but would slightly affect the situational statistics of hitting? Most of these mistakes discovered at year-end were in the minors—multiple layers of checking had already been done for the major leagues—but nevertheless, the operation worked to ensure the data were as clean as possible.

Over time, increased data collection efforts have been layered "on top" of the existing database. When MLBAM bought the Total Sports system in 2000, it effectively bought the relational database. Every person associated with that database—whether a high school amateur, official scorer, minor leaguer, data stringer, coach, executive, or umpire—was given a "primary key" that she or he keeps forever so that a person's complete participation across multiple roles can be called up quickly. Likewise, IDs are assigned to media figures and celebrities (to track their engagement with MLB on social media), teams, and venues. More impressively, the span between each pitch also has a generated unique ID, or GUID. Everything that happens between pitches is identified by this GUID, so that it can be recalled with a simple search.

On top of all this, the new Statcast system overlays its data on these GUIDs. Statcast is a set of tools that effectively tracks the precise location of every player and umpire, as well as the first and third base coaches, at every moment, alongside the movement of every ball. Two technologies to accomplish this were used originally: one, developed by a company called TrackMan, was Doppler-based and could track the speed, angles, and movement of balls, while the other, which was video-based, was developed by a company called Chyron-Hego and could track both balls (with lower fidelity) and bodies.

Now, in any particular span between pitches, the associated GUID is linked to a series of player movements, as well as a series of arcs of the moving ball, which includes the physical properties as the ball moves from a pitcher to a catcher, off a bat to the stands, or from one player to another. Each of these arcs also receives an ID, so that for any particular GUID, there is a "play"—perhaps a single to right field—a set of motions of the players, and a set of arcs (the pitch, the single to right, the throw in from the outfield) prior to the next pitch. Of course, all of these are subject to measurement error, but the technology is believed to provide center of mass calculations within inches, close enough to know where the person or the ball is for practical purposes.

This is the brave new world of big data. It's now possible—or soon will be—to pull up every relay throw a particular right fielder has made during Saturday night games in Cleveland and compare their speed, trajectory, and accuracy. Indeed, after the 2016 season, the second in which MLBAM provided detailed data to the clubs, a revolution was promised in news coverage in nearly the same language used in *Moneyball*: "There will be an upheaval in the way ballplayers are valued, from roster decisions to salary structure to postseason awards."[26] The new data promised to reveal which of the seemingly ordinary players were in fact future stars, just as the data gleaned from Project Scoresheet and STATS and analyzed by Bill James enabled Billy Beane to cobble together a winning club by knowing which statistics were important.

These comparisons raise the question of whether such big data represent a difference of degree or of kind. Tom Tippett and Dick Cramer's decisions in the early 1980s about how to simulate and record the game on electronic computers and Gary Gillette, John Dewan, and Dave Smith's experiences collecting real-time data are vestiges lurking beneath the fancy graphics of exit velocity and pitch angle that appear on the nightly broadcasts. The whole thing is built upon a structure that dictates how to turn the game into written inscriptions that can be recorded, compiled as code, and then displayed in a useful way. A version of the code designed in two weeks' time to allow volunteers to track games in the 1988 season was being sent nearly three decades later to every major league club each morning, giving them the "official" play-by-play of the previous day's games.

Over time, the collection of data has gotten easier. Now, data are "born digital" in the computers of MLBAM rather than re-corded by hand, input into digital memory, and then aggregated.[27] That hasn't eliminated the need to authenticate and verify the data, however, and those systems remain in place.

Nevertheless, it is clear that the petabytes of Statcast data avail-able are qualitatively different than the bound volumes Project Scoresheet produced. Even though electronic computer process-ing is often based explicitly on previous nonelectronic ways of gathering data—just as the DiamondWare system was constructed

TOP OF THE 1st

AB#	Play String	Play	Play by Play
1	.CBX	7/F	Brett Gardner flies out to left fielder Joey Butler.
2	.CFS	K	Carlos Beltran strikes out swinging.
3	.BBCE	HR/78/F	Alex Rodriguez homers (8) on a fly ball to left center field.
4	.CSBFX	5/L-	Mark Teixeira lines out softly to third baseman Evan Longoria.

FIGURE 4.4: Through 2015, Major League Baseball Advanced Media provided clubs play-by-play accounts using the original inputting language of Project Scoresheet for both pitches ("play string") and plate appearances ("play"). Courtesy of MLBAM

to incorporate Wright's scoresheets—large-scale data operations may generate new questions and practices. Sportswriter Keith Law compared the situation to physics: "The old data got us to the atomic level, but Statcast data gets us to subatomic particles we couldn't measure before the new technology arrived." The claim that scaled-up data transforms how we see the world is not limited to baseball enthusiasts. Historian of science Hallam Stevens also noted a similar effect, in that changing the scale from solid-state physics to particle physics *created* new phenomena—it wasn't simply a matter of "scaling up." "At some point," Stevens concluded, "more data is different."[28] With increasing scale there can be qualitatively new observations, new questions, and new theories to investigate. Statcast uses a mix of old and new forms of data collection to enable qualitatively different phenomena to be revealed.

Statcast's history suggests that the rhetoric of revolutions or sharp breaks with past baseball data-gathering efforts should be moderated. The analogy with physics again proves illuminating: instruments, observations, and theories need not progress in any fixed or predetermined way. In physics, no less than in baseball data collection, new insights may be fashioned from old ways of observing, just as new instruments may confirm familiar insights.[29] It seems easy to divide up the world into an old era of counting home runs and a new one of exit velocity, launch angle, and hang time, but historically such stark divisions never exist. Tools, concepts, and expertise easily slip from one side of the divide to another. MLBAM's Statcast still yields the numbers for those who care about traditional statistics—hits, errors, runs—alongside its new measurements. The system remains an outgrowth of efforts to translate a complex game into a series of discrete events. Statcast was built on the infrastructure of Project Scoresheet–like efforts to record baseball as a series of discrete inputs.

Data operations like Statcast, at first glance the antithesis of subjective, human-based knowledge, continue to depend on the human labor and expertise needed to turn baseball into a manageable set of data. Nearly everyone involved in the design of play-by-play systems had extensive training in mathematical or scientific fields. There is no clear technical–human divide. Dave Smith, despite his academic training in science, spends a great deal of time making historical judgments about the reliability of sources. Cory Schwartz, despite his title as vice president of statistics, spends most of his time managing people rather than fiddling around with analytical techniques or troubleshooting technologies.

It is easy to parody modern analytics and overly specific statistics about first-step efficiency during Thursday afternoon games in May. But whether a fan or a critic, it is impressive that we have statistics with this level of detail at all. It took a major infrastructure project many years to present data as seamlessly emerging from the game, as if they were just out there, waiting to be displayed.

5

The Practice of
Pricing the Body

Though few predicted Craig Biggio would be a Major League All-Star and even fewer bet he would have a Hall of Fame–worthy career, he was widely acknowledged as a top prospect after record-setting seasons as Seton Hall's catcher from 1985 to 1987. His playing statistics were exceptionally good and still rank in the top 10 in Seton Hall history in 18 single-season and career statistical categories. But there are many college stars who fail to become major league players.[1] Even top college prospects are gambles for professional clubs, bets that have only a relatively small chance of paying off.

That's why scouts are essential to most clubs. They are in the business of determining a player's ability to produce value in the future. The Chicago Cubs, for example, wanted to know if they should draft Biggio, so they sent their veteran scout, Billy Blitzer, to watch him in the spring of 1987. That June Biggio would be eligible to sign a professional contract, and they wanted to know how much he might be worth. The report Blitzer subsequently produced was extensive, with descriptions of his body ("muscular,"

CHICAGO CUBS
FREE AGENT REPORT

SCOUT _B. Blitzer_

UNIFORM NO.	GROUP NUMBER	DOLLAR VALUE	DATE	REPORT NO.
44	1	$65,000	3/17/87	1

	LAST NAME	FIRST NAME	MIDDLE NAME	POSITION
PLAYER	Biggio	Craig	Alan	C

CURRENT ADDRESS	STREET		TELEPHONE (AREA CODE)
			( )

CITY	STATE	ZIP CODE	DATE OF BIRTH	HEIGHT	WEIGHT	BATS	THROWS
South Orange	New Jersey	07079	12/19/65	5'11	185	R	R

PERMANENT ADDRESS (IF DIFFERENT FROM ABOVE) — Kings Park New York 11754   GRADUATION 6/88

SCHOOL	CITY	STATE
Seton Hall University	South Orange	New Jersey

COACH	TELEPHONE (AREA CODE)	GAMES	INNINGS
Mike Sheppard	(5)201-761-9196	2	15

SUMMER CLUB	CITY/STATE	60 YD. TIME
Cape Cod League - Yarmouth-Dennis	Cape Cod, MASS	

RATING KEY	NON—PITCHERS	PRES.	FUT.	PITCHERS	PRES.	FUT.	USE WORD DESCRIPTION
0 - Outstanding	Hitting Ability	3	5	Fast Ball			Habits _Good_
9 - Very Good	Power	3	4	Curve			Dedication _Good_
8 - Well Above Average	Running Speed	6	6	Control			Agility _Exel_
7 - Above Average	Base Running	5	7	Change of Pace			Aptitude _Good_
6 - High Average M/L	Arm Strength	5	6	Slider			Phys. Maturity _Good_
5 - Low Average M/L	Arm Accuracy	5	6	Knuckle Ball			Emot. Maturity _Good_
4 - Below Average	Fielding	4	6	Other			Married _No_
3 - Well Below Average	Range	5	6	Poise			
2 - Fair	Baseball Instinct	5	6	Baseball Instinct			Date eligible 6/87
1 - Poor	Aggressiveness	5	6	Aggressiveness			
	PULL Str. Away Opp. Field			Arm Action			Phase _Regular_
	X			Delivery			

Physical Description (Injuries, Glasses, etc.)

Medium frame, stocky, compact appearance. Muscular body with no fat. Thick legs and upper body. Weight lifters appearence. Although body is muscular it has maintained exel bounce and agility. No glasses or contacts. No known injuries.

Abilities Soft, sure handed receiver who shifts well, blocks balls and has quick feet. Short quick snappy throw with c arm and accuracy. Plus speed - 4.15 to 1st - Aggressive base runner capable of stealing bases. Fluid running stride. Hits out of straight up stance with slight crouch. Relaxed compact swing. Line drive gap hitter. Uses all fields. Will lay down bunt and uses speed. Good eye. Patient hitter. Tough aggressive kid.

Weaknesses

Only catching 3 years. Tends to snatch at pitches. Hitting style lacks power. Makes contact and runs.

Summation and Signability

Biggio has had fine defensive skills and speed but his bat and other areas of his game have shown great improvement. Versatile athlete who can play infield and outfield besides catching.

MI -18

with "no fat" and "thick legs"), possible impairments ("no glasses or contacts" but "lacks power"), his mental state ("good" aptitude, "good" maturity), and his future ("versatile athlete who can play infield and outfield besides catching").

The best scouts are experts in evaluating amateur prospects, with time-tested predictive heuristics and detailed systems of evaluation. They have very specific ideas about how to watch players, how to transcribe what they see into language useful for clubs, and ultimately how to translate those judgments into a dollar amount or simple number representing a player's value. Ultimately, Blitzer put a value of $65,000 on Biggio; the Cubs instead drafted pitcher Mike Harkey and signed him for $160,000.[2] Harkey had a superb rookie season in 1990 followed by a mediocre career. Evaluating a prospect is never easy.

Scouts may describe themselves as hunters of talent, but what they actually do is create a paper record of their attempts to precisely evaluate prospects. They write reports. Longtime scout Gib Bodet lamented the never-ending drudgery of this task. He claimed it was "terribly tough to keep writing reports, especially when you're on a heavy traveling schedule. I write a lot of reports in airports, a lot of them. I write a lot of reports when I'm home, of course." Another longtime scout, Hugh Alexander, told Kevin Kerrane in the early 1980s, "Most scouts don't romance prospects anymore; they just write up a shitload of evaluations." Or, more politely, old-timer Charlie Wagner said: "The young scouts are better organized than we were, because so much of scouting now is writing reports."[3]

On the surface, written reports are the physical residue of scouting, documents usually seen by only a couple of people before they are filed away or thrown out. But they are critical to scouting, constituting a highly refined and formalized technology of communication. Not just any information is written down. Through reports, the subjective nature of scouting—the process of one human watching and evaluating another—is made bureaucratic, formal, and numerical. Indeed, just a quick glance shows how important numbers were for Blitzer's report: he indicated height and weight, running speed, and innings watched, and he provided numerical measures of Biggio's skills. Scouts have to learn how to give numbers to present and future assessments of factors such as "power" or "baseball instinct." They have to learn how to place a dollar value on a player.

Reports are essential tools of bureaucracy. Indeed, though Blitzer worked for the Cubs and learned his trade from one of the scouting greats, Ralph DiLullo, he began his career as a part-time employee of the Major League Scouting Bureau, an organization whose dedication to bureaucracy was so thorough its directors put the root word in its name.[4] Formed by teams to save money by pooling scouting resources, the bureau for many years ran a "scout school" to produce people like Blitzer who were able to translate players' abilities into numbers and their bodies into dollar values.

The role of reports and bureaucracy in the quantification of baseball prospects is a story that has long been obscured by a romantic notion of what scouts do and who they are. Outside of scouting memoirs, only a handful of book-length studies of scouts exist, none of which take scouting tools and training as the central topic. Scouts actively participate in their own mischaracterization. It's possible to read entire memoirs of scouts without ever learning about the need to fill in a report, let alone how it is done. Scouts instead describe themselves as "baseball men" dedicated to figuring out who can really play the game. Interviews with scouts, particularly those conducted by Kevin Kerrane, P. J. Dragseth, and the SABR Scouts Research Committee, nevertheless reveal the ways in which written reports, and the tools that are required to produce them, are essential to making knowledge about prospects and their worth useful to clubs.[5] The scouting report is a crucial step in turning subjective judgments into reliable, useful knowledge. We know that it is useful in part because scouts have survived the last four decades of statistical analysis in baseball. They're still filling out reports detailing what they see.

This chapter, along with the next two, shifts from the history of scoring to the history of scouting. Like scorers, scouts have been concerned with making their judgments reliable and precise. There is great financial incentive to sign high-quality players early. Not only can clubs extract their labor cheaply, but they can also sell the player later at a premium. Professional baseball clubs have a huge stake in the accuracy of scouts' judgments, and chapter 6 will detail how they have implemented regimens of management and control to govern these judgments, strategies that draw on the human sciences more broadly. Scouts have attempted to measure physical ability in fine detail, to produce objective knowledge about mental states and cognitive characteristics, and to find reliable ways of using past performance to predict future value.

Amateur scouts themselves tend to divide their profession into pre- and postdraft eras, splitting its history with the inauguration of the major league draft in 1965. This event was indeed important, and chapter 7 will show how it resulted in concrete changes in scouting, especially in the use of highly quantified reports. But the core job requirement of amateur scouts hasn't changed in over a century. They need to be able to accurately evaluate and price the worth of the future labor of a prospect who can only be seen in the present. Like scorers, scouts attempt to translate the messy world of young people playing ball into marks on a page that can be mobilized as reliable data. Amateur scouting is a scientific practice of evaluating bodies.

———

First appearing regularly in the early twentieth century, the word "scout" was used as shorthand for those who were searching for talent to acquire. As early as 1910, sportswriter Hugh Fullerton and second baseman Johnny Evers wrote in their book *Touching Second* that the supply of top-caliber players was so small that "the owners of clubs in the American and National Leagues and the higher minor leagues have resorted to dragnet methods to discover them." By this point each club had a "scouting force" of veteran players or retired owners whose duty was to "scour the entire country," "seeking men who by their playing show promise." Clubs received "hundreds of letters" each week proclaiming the skill of various players, the most promising of whom the club might dispatch a scout to see and report back on. Fullerton and Evers gave the example of Barney Dreyfuss, owner of the Pittsburgh club, who kept books in his office with prospects' names, addresses, descriptions, batting and fielding averages, and "character and general makeup." Most of these prospects, everyone knew, would never play a day in the major leagues.[6]

Each club generally operated independently in these early years, meaning that signing a player meant putting him on the roster. There was effectively no "farm" system, and for much of the late nineteenth and early twentieth centuries, the "major" clubs (at least the richer ones) could just wait to see how players performed elsewhere and then buy out their contract. Major league scouts might recommend a player to a minor league club in exchange for the option to buy back the player later. At the same time, minor league clubs had their own scouts looking for players who not only might improve their team but also might develop into an asset worth selling to a richer club at a profit. The whole business was in some respects smaller than the present one; prior to World War I, personnel decisions were handled directly by owners and managers rather than by an assortment of general managers, farm directors, and others. Scouts employed by a major league club might maintain various other connections with minor league teams and baseball operations, ensuring they profited both from finding the player and from developing him. Larry Sutton, who became one of the first full-time scouts for the Brooklyn club in 1909, was conveniently also the business manager of the Newark minor league club. Cy Slapnicka, a pioneering scout for the Indians, also briefly managed minor league teams and worked in Cleveland's front office.[7]

The practice of scouting was built upon rules concerning labor contracts between players and clubs. There were few restrictions on how teams might sign players in the early years, and common metaphors of the time reflected this freedom. Scouts were called "bird-dogs" or "ivory hunters" as if they were searching for big game, assisting teams in the hunt for trophies. Though distinctions were sometimes blurry, "bird-dog" was often used for part-time scouts who provided tips on players, whereas "ivory hunter" was used for official representatives of teams.[8] Either way, the reference was to hunting game, and, indeed, when a scout found a target he liked there were few restrictions on "bagging" him.

Scouts remembered this interwar period fondly. Hugh Alexander started scouting in 1938 and recalled much later that "it was so easy in those days. There were so many ballplayers and so few scouts." Of course, the ability to tap into a steady stream of cheap prospects also depended on a club's finances. Clubs could sign players only when they were able to agree on a price for the prospects' labor. Not surprisingly, the wealthy New York Giants and Brooklyn Dodgers had two of the most extensive scouting networks in the nation, in no small part because there was no sense in having a scouting network if a club couldn't afford to sign new players once they were found. The Yankees under de facto general manager Ed Barrow had particular success at signing top amateur players to ensure the major league club was well stocked for decades. If the Yankees wanted a player, they wouldn't lose him over something so minor as money. Nevertheless, even wealthy New York clubs were not major operations. The Yankees' scouting department through the 1930s had fewer than a dozen full-time members.[9]

Cy Slapnicka's signing of Iowa farm boy Bob Feller for $1 and an autographed ball in 1935 is often taken as the model of this era: an unknown prospect signed for almost nothing by an enterprising scout. Not surprisingly, the story wasn't really that simple. Feller was hardly an unknown, at least to those who had seen him play, and the case of his signing was typical in a different sense—it was a subterfuge to control his future labor as cheaply as possible. Slapnicka had signed Feller to a minor league contract while he was still in high school to protect him from other teams and was consequently able to transfer Feller seamlessly (and cheaply) to the Cleveland club when he was eligible to sign a major league contract after graduation. In effect, Slapnicka was able to bring Feller to the majors through a minor league contract without needing him to play once at that level. The system worked perfectly, at least in the sense that Slapnicka scored a huge talent at low cost.[10]

The hunting metaphor for scouting remained appropriate in another sense through the Second World War: it was a dangerous jungle out there for prospects. Unsigned players were largely free to commit to any team at any price, but once they did so, they lost nearly all control over their labor. Baseball's infamous "reserve clause" ensured that clubs could "reserve" the labor of their players at the expiration of each contract, retaining the ability to re-sign, trade, or sell them. Though the number of players that could be reserved (and the mechanisms for reserving them) fluctuated for nearly a century after the mechanism's creation in the 1870s, the reserve system worked effectively. Prospects had one moment of control—before signing their first contract—and then the leverage of all but the superstars was gone.

This system came under periodic challenge throughout its existence, but its most stable period was from the 1920s to the 1950s, the generation after Supreme Court Justice Oliver Wendell Holmes Jr. penned the decision in 1922 affirming that baseball was not "interstate commerce" and that its reserve clause was therefore constitutional.[11] This was, not coincidentally, the period in which "farm" systems developed to ensure that as soon as a ballplayer signed his first professional contract, the clubs—not the player—would control his labor as his talents developed.

Until the Depression, such control was exerted from a distance, because minor league clubs were often independent organizations. That situation changed as economic troubles brought more minor league teams under the control of major league clubs. Once players were in the minor leagues, clubs could use the leagues to develop players and to determine who would be worth promoting. The goal was to sign a great number of prospects to cheap contracts, see who developed into good ballplayers, and then either promote them or sell them at a premium price. This was longtime executive Branch Rickey's system of "ripening" players into money to produce "quality out of quantity." The more players signed to his ever-growing network of minor league teams, the more talent was available cheaply for the primary club. This system depended upon the steady work of amateur scouting, and indeed, two of the most prolific scouts in this period worked for Rickey's St. Louis Cardinals, Pop Kelchner and Charley Barrett.[12] By 1940, the Cardinals controlled 33 minor league teams, about 1 out of every 10 in the country. Other wealthy clubs enjoyed similar riches. Rickey's counterpart for the Yankees, George Weiss, had 15 teams under the Yankees' control by the mid-1930s. Likewise, the Dodgers had 29 teams by 1947.[13]

Though this system worked well for wealthy clubs, not everyone in baseball thought it a success. Commissioner Kenesaw Mountain Landis wanted to keep the minor leagues from becoming controlled fiefdoms and so pushed for expanding the minor league draft, in which rival clubs could essentially pick off minor league players not on the 40-man roster.[14] For scouts of wealthy teams, the decades leading up to midcentury presented the opportunity to focus on signing players with relatively little risk.

As the minor leagues contracted in the 1950s, there were fewer roster spots in which to stockpile players and await their "ripening." Between 1949 and 1956, 31 leagues and 240 teams went out of business, and over the next six years, major league club owners created funds to stabilize teams and eventually reclassified most minor league teams as AAA, AA, and A affiliates of the parent club. The arrangement was again ultimately about control of players' labor, with major league clubs owning the players and minor league clubs owning the stadium and profiting from ticket and concession sales.[15]

These economic changes precipitated changes in the practices of scouting. The number of major league clubs by the 1950s (16) was the same as in 1903, and yet far fewer other clubs existed for which players could play professionally. For postwar scouts, this meant more competition to get top players into their system. More clubs were by this time also following the example of the Cardinals and the Yankees, expanding their scouting departments to try to locate more prospects. Cincinnati's club had no full-time scouts in 1935 but nearly 40 by 1959. Such rapid growth was typical. On average, there were only about 4 full-time scouts per club in 1941, about 15 in 1949, and nearly 30 in 1965.[16]

The increased numbers of scouts in this period, coupled with the diminishing number of roster spots, meant a lot of competition for top prospects. For clubs this often resulted in ever-increasing bonuses paid to the players—a factor that would eventually be decisive in ushering in the draft. For scouts, this meant an increase in the use of unsavory means to sign players. "Taking shortcuts was more the norm than not," as Slapnicka would put it.[17]

The era of the 1950s was therefore one of increased stature and importance for scouts. Those like Randy Gumpert who worked for wealthy teams could essentially sign whoever they wanted; Gumpert was told by the New York Yankees that a scout's goal was to find and sign a "prospective super player," no matter the expense. If a supervisor wouldn't "go as high as . . . he should" in producing a bonus payment, the scout had to "make him" pay up for the sake of securing the prospect. For most clubs, however, money remained the limiting reagent: front offices didn't want scouts overpaying for prospects. Nevertheless, scouts still retained a great deal of control. Jerry Gardner, who became a full-time scout for Pittsburgh in the 1950s, recalled that unless a player was still in high school, "it was more or less that we could sign these fellas." Basically, he said, "we just sent a report in" on a prospect, "called the office on the phone to tell them what we thought of the guy and how much it would take. And he would give us a yes or no." Usually this entailed the front office replying, "We can't give him that much money. Get him for this amount." As Gardner said, "That was the way we operated in those days."[18] Teams were limited only by roster space and money.

Scouts had to be effective salesmen in this era, selling not just the profession of baseball but also their particular organization and its advantages to prospects. Scouts would pursue parents aggressively, pushing their own organization as the best fit for the child or as the best chance of his playing at his top potential. Gene Thompson (another five-decade scout) recalled that in the 1950s scouts would follow a handful of top players, get to know their families, and try to use that information to sign them. When Fred McAlister started scouting for St. Louis in the early 1960s, he was told he had to "court" the families of top players. The Yankees encouraged scouts in the 1950s to cultivate the right people and become a real friend to prospects—in short, to "be friendly" at all costs. For scouts, in an era when just about any amateur of a certain age was available for the right price, personal relationships were key. Indeed, scouting advice and lore from this era focused largely on personal networks. When Slapnicka gave advice to the young Hugh Alexander after the latter started as a scout in 1938, his words weren't that far off from what could be found in a guide to good manners: "Make only promises that you can keep," "Know the player's mama," and so on. Finding players was a matter of finding "networks that know talent" just as much as it was a matter of evaluating the players. Knowing people—in both senses of the phrase—mattered.[19]

The people business was, however, still a business, with all its attendant needs: paper trails, elaborate chains of command, and bureaucratic demands. Even in this "informal" period of scouting, when one scout could find an eligible amateur, get to know him and his family, and then sign him, there was almost always a paper trail. Harry O'Donnell, working as chief scout for the Cleveland Indians in the 1950s, would write regularly to his then-boss, Hank Greenberg. Sometimes O'Donnell would report on the prospects he'd seen or tryouts he'd held; sometimes he would ask permission to sign a player; sometimes he would send evaluations back to the front office. O'Donnell was essentially running a correspondence network, with a side business of signing players.[20]

Some clubs would address their business needs in the time-honored fashion of creating manuals. One that has been preserved is the 1949 "Operations Manual" of the Pittsburgh National League club, the Pirates. Presented by General Manager H. Roy Hamey and Secretary of Farm Clubs Fred Hering to the club's 14 or so full-time scouts, the manual specified the organization's operations in an attempt to prevent scouts from "making unnecessary errors through administration of the business end of baseball." The goal, of course, was to sign players to the cheapest contract possible without losing any player that a scout liked. Though the scout still had wide discretion in signing players, any expensive contracts had to be immediately reported to the Pittsburgh office.[21] There was incentive on both sides to find "quality material" at low prices.

The manual was not overly focused on *how* to evaluate players, however. Each scout was presumed to be autonomous in this area, at least in 1949. Rather, the bulk of the manual was focused on diligent recordkeeping. Written records were essential and prized over verbal communication. The Pirates' front office exhorted, "Don't phone when you can wire. Don't wire when you can write." A mnemonic was suggested: "B—for Baseball: BE prompt in answering mail. BE efficient. BE conscientious."[22] Scouts may have been valued for their ability to spot and sign talent, but from an organizational standpoint they needed to be good bureaucrats too.

The operations rules took up the bulk of the recommendations. Bulletins from the front office should be placed in the front of the manual. All reports should be signed and dated. Expense reports should be filled out in triplicate and mailed to the front office on the sixteenth and last days of each month. Itinerary cards had to be mailed to the office every Friday. All major colleges in a scout's territory had to be specified and reports filed on each one's players.[23] Good scouting meant good paperwork.

Such recordkeeping, from an organizational standpoint, was essential. This applied not just to expense reports, but also to the reports on players themselves. Herb Stein—who had started scouting for the Washington Nationals in 1952 and ended his career in

1994 with the same club, then in Minnesota—scouted a young Rod Carew in the early 1960s when he attended a game at Stein's alma mater, George Washington High School. The account of the signing echoes the classic "diamond in the rough" story: Carew, a Panamanian, was not eligible to play interscholastic baseball in high school, but Stein heard about him from his network of "bird-dogs," who happened to catch Carew playing on a sandlot team in the Bronx. On the night of Carew's graduation—he'd be eligible to sign at midnight—Stein took Carew out to dinner to ensure he had first access to him at the stroke of midnight. When Carew wanted to think the offer over, Stein gave him the phone number of a rival scout who Stein knew would be unable to offer something comparable right away.[24] Stein indeed got his man that night.

The signing of Carew was also indicative of scouting's bureaucratic nature by the 1960s, however, even if it was not often portrayed that way. Stein had filed a "Player Information Card" on Carew earlier in 1964 designating Carew a prospect for the organization. An "X" for "excellent" was marked to note the scout's evaluation of Carew's arm strength and accuracy, fielding and hitting ability, hitting power, and base-running speed. The card also indicated that by May 1964, Stein had seen him personally in four games and described him physically as "good all the way." Stein believed Carew had a chance at becoming a professional success, though at this point he was not yet a "definite prospect." The report also gave credit to Monroe Katz, associate scout for the Twins, for recommending Carew to Stein. Stein further reported Carew's parents' names and his high school, basic biographic information, military and free-agent status, and recommended classification after signing. The card was a way to track both Carew's career and Stein's business activities and processes.[25]

Just before the signing, Stein filed a more extensive report on Carew with the front office, this one in the vein of more formal letters like O'Donnell's. In it he began by noting that he had followed Carew for "a number of games" between April and June 1964, and that Carew would be graduating (and thus eligible to sign) that summer. He then presented a series of evaluations of Carew himself: "Physically he has a fine body, wiry type. He has no known injuries or physical defects. He has good hands, reflexes and range. His movements and actions are good. He is knowledgeable and has shown good baseball instincts, which is rare among young players. . . . He is an excellent runner. . . . He is a good hitter." His weaknesses (arm strength, double play foot motion, ability to "get rid of ball quicker") were noted, but so was his character as a "quiet and nonchalant player" who "does the job." Finally, Stein bluntly evaluated his prospects: "Once in pro ball[,] if he can get use[d] to being away from home, accepting the good and bad, and the many adversities in this game, he will become one hell of a player."[26]

There was a great deal that had to be made precise. Carew's height and weight were prominently listed, and Stein noted that Carew generally made it to first base in 4 seconds and could do 3.8 seconds on a bunt. To justify his claim that Carew had "shown good power," Stein noted that he had personally seen him hit balls 400 feet. A player's age was essential to get right—not just so that there would be an accurate sense of how many years were left to develop his talent, but, more importantly, so that the exact date of his eligibility to sign was known. This wasn't a "trust me" letter—it was one that clearly was making a case for why this particular player would be worth a specific investment.

The letter also mentioned that not many other clubs know about Carew, despite his "great potential," and ended by noting that Stein was prepared to offer Carew a "cash bonus" of $5,000, an incentive bonus plan (an additional $2,500 if he made the major league roster), and $400 per month. "My relationship with him is good, and [I] have [a] good chance of signing him," Stein noted. A positive response to such a letter would be a prerequisite to Stein cornering Carew the night of his graduation. The signing did involve some scheming on Stein's part, but only at the end. The bulk of the process involved mundane paperwork.

Stein's report cards and O'Donnell's letters were increasingly typical practices for scouts over the 1950s. Joe Stephenson, a

PROSPECT REPORT — MAKE 3 COMPLETE COPIES: ONE FOR BOSTON MINOR LEAGUE OFFICE, ONE FOR CLUB, ONE FOR SCOUT WHO SIGNED PLAYER.

NAME: Mike McCormick (OUTSTANDING)
RECOMMENDED BY:— Joe Stephenson DATE: 11/1955
STREET:
CITY: Alhambra STATE: Calif
POSITION: 1B-P BATS: L THROWS: L HEIGHT: 6'1" WEIGHT: 170
DATE OF BIRTH: (9/29/1938)
SCHOOL & DATE GRADUATE:— Mark Keppel High–June, 1956

BONUS ARRANGEMENTS: (WORD EXACTLY AS IN CONTRACT)
Bonus player $50,000 New York Giants Aug. 1956

DETAILS ON PLAYER:—

Player is strong LHP with real good speed and an outstanding curve ball. Has showed in games that I have seen him that he's a good hitter. Can hit the long ball. Is a little below the average in running. Is an excellent competitor. Good clean cut boy. Should start in C or D ball with a chance to go all the way.

FIGURE 5.2: Joe Stephenson's 1955 report on Mike McCormick functions both as a business record and as an accounting of the prospect's value. Courtesy of the Boston Red Sox, Scouting Report Collection, BHOF

longtime scout for the Red Sox, filed reports during this era that epitomized this need to quickly describe players and their value. The forms saved a third of the space for the player's name, address, high school, date of birth, and handedness; a third for contract information; and a final third for "details." Stephenson specified characteristics like "outstanding," "good habits," "excellent speed," or "quick to learn." But he also included relevant snippets to give a sense of the person: "Is an excellent competitor. Good clean cut boy."[27] Many of his reports indicated how long he had known the player, and in most cases he had followed the prospect for a while, sometimes throughout high school. The brevity of Stephenson's reports is in some ways an indicator of his familiarity with the players; these were simply the essentials passed along as a report to the front office. (Stephenson's reports were filed in triplicate, one for him, one for the minor league office, and one for the major league club.) He didn't need to convince higher-ups that a player would be a star, just that a player was worth signing. Stephenson later annotated his typed reports with the bonus amount agreed to by players and clubs, even if another club signed the prospect. They were as much "expense" reports as scouting reports.

———

Scouts don't often talk about these reports. Interviews with scouts, as well as their memoirs, all cover such well-worn ground that they sometimes read as if they were a parody of the genre. The marginal or unlucky player sticks with the game as a scout and begins to hunt the backwoods and sandlots for hidden talent. He finds a few key players who blossom into major league stars and becomes life-long friends with them. Along the way he has lots of misadventures in bad motels and beat-up cars but ultimately is able to outwit other scouts and find the diamonds in the rough. This emphasis on the *rough* is essential: scouting isn't just about evaluating the diamonds everyone else knows are valuable. It's about finding them in the least likely of places, ideally with a good story to go along with the discovery. Nearly every veteran scout has this kind of tale at hand, a time when he was headed out on back roads or through dark alleys to hunt for the players no one else knew about. The bigger the gap between a player's initial obscurity and later fame, the better. Scouts have tended to describe themselves like cowboys or long-haul truckers—the misfits and loners of the world. Like cowboys and truckers, however, scouts are actually valued components of large, bureaucratic, and often financially sophisticated enterprises, even if they also are underappreciated old men with an "eye for talent."[28]

Scouts have historically been measured by the quantity and quality of major league players they sign. Even among themselves, scouts have largely described the job this way. In 1949, the Pirates' "Operations Manual" praised scouts as the "life line" for securing players for the organization and wished them "good hunting." A half-century later, the scouting manual for the Texas Rangers explained their role in nearly the same terms: "Each day you must say to yourself, 'Today is the day I am going to find a Major League Player!'"[29] Even the Society for American Baseball Research lists scouts by the professional players they've signed; this is just the way they're known.

Had the Cubs signed Craig Biggio, Blitzer would have gotten "credit" for him, regardless of who noticed Biggio first or whether others had rated him more precisely or accurately. Some longtime scouts, like Hep Cronin, complained about this association:

> It's amazing the respect you get once some of your players begin to perform. You happen to get lucky one or two times and all of a sudden people will say, "Oh, this is Hep Cronin. He signed Dave Justice." It makes me feel a little goofy when I hear that. It doesn't take a genius to see that a player like Dave Justice has great potential.

These are the "easy" ones, said Cronin. But the really skilled scout is the "guy who discovers someone the others missed or underrated."[30]

The titles of scouts' own accounts are telling—nearly all focus not on their own education or their own view of the game, but rather on their attempts to find players: *Old Baseball Scout and His Players* or *Diamonds in the Rough* or *Journeys with a Major League Scout* or "Baseball's Road Scholar." These books are nearly all organized as a series of anecdotes, one chapter per adventure. They are often written with the perspective that Tony Lucadello brought to his own account. Telling his biographer, Mark Winegardner, to take a look at the list of players he had signed, he declared, "Pick any one of those players and I'll tell you a story. There's a story that goes with every player I ever signed." Winegardner himself speculated that perhaps these stories arise not because they're particularly telling or important, but rather because there's a lot of time to kill in baseball, and those who stay in the game for years have plenty of opportunity to refine their storytelling skills.[31]

PRICING THE BODY 153

Scouts have traditionally written about themselves in a manner that effaces the sort of people they actually were. In the past, the two most common compliments among scouts were that they had been doing it forever—and therefore were entitled to be called "Ancient Mariner" or "Old Scout"—or that they were "good baseball men." The idea of a "baseball man" is a telling one—he was a man, for one thing, but a man whose main attribute was the sport itself. Longtime scout Jocko Collins described a baseball man as one for whom "there's no line anymore between him and the game, no point where baseball leaves off and he begins."[32] Scouts could be old because age was seen as a signifier of experience and knowledge. (Scouts tellingly called prospects of all ages "boys.") The older they were, the more they were assumed to know about the game, or at least the wilier they were thought to be.

The value placed on experience in scouting is evident in the complaints some scouts had about the younger generations. Cecil Espy, who scouted for over two decades, lamented,

> Take a look at scouts now, all young college kids who probably never played or even seen minor league baseball. They don't have the background of the game. . . . It used to be that all the scouts and the front office personnel had a background in the game and we had baseball in our blood, as they used to say.

Longtime scout and scouting director Fred McAlister said that without experience, "it's a part of their education they've missed and there's no way to make it up." Similarly, Jethro McIntyre felt new scouts operated out of a "book." Phil Rizzo put the same point more colorfully: "Kids today don't know nothing about the game." His reasoning? "Guys who want to scout should at least have played the game to have an idea what a ballplayer goes through. They have to know about what obstacles you have to walk through before you become a player." Nowadays, "They can read books and they think they're learning how to scout. Heck, many of the scouts in my day probably didn't even know how to read."[33]

Among scouts, however, there wasn't uniform agreement on the point of whether a scout should have previously been a player.

Other scouts were more like Julian Mock, who believed that "guys who haven't played the game do become scouts and some are good at it because they have athletic instincts"—meaning that they were able to evaluate bodies and how they might develop. Moreover, others claimed that having been around the game for too long might cause blind spots as a result of groupthink. Jim McLaughlin, Baltimore's onetime scouting director, declared that he wasn't a "real baseball man," because "baseball men are like a tribe," and he preferred to take a viewpoint that sometimes went against the grain.[34]

Not all scouts were formerly great ballplayers—few were—but their comments all pointed to a belief that the skills involved in scouting were not teachable, and certainly not acquirable through books. Rather, scouts had to follow the game—live the game— over many years to know what they were doing. Scouts certainly needed to love what they were doing, because they weren't likely to get paid extravagantly for it. Over the years, part-timers and other beginners have typically received only a couple hundred dollars for recommending players who worked out. Full-time scouts rarely made even a small percentage of front office or player salaries—the respected veteran scout Joe Branzell made only $8,000 per year in 1971 and $32,000 in 1986 (about $70,000 in today's money).[35]

What no one talked about were the scouts' own bodies. They were male, almost by assumption, though there have been a few husband-and-wife teams that have scouted together and one well-known female scout, Edith Houghton. A few female scouts have been recently hired, and there are almost certainly more women involved with scouting behind the scenes, but the fact of the matter is that scouts historically were assumed to be not just male, but aggressively male: cigars, wide-brim hats, sizable bellies, and penchants for fast food.[36]

Though scouts once were exclusively white and are still a fairly white bunch, there have been a number of successful scouts of color, following John Wesley Donaldson's debut as a black scout

PRICING THE BODY **155**

for the White Sox in 1949. A 1967 issue of *Ebony* noted that there were only eight full-time black scouts nearly two decades after Donaldson entered the industry. The experience of black scouts is a good reminder of the ways scouts' bodies *did* in fact matter. Buck O'Neil felt that his blackness gave him a "tremendous advantage," in that coaches and managers of black teams would share information with him that they wouldn't tell a white scout. Ellis Clary formed an "underground scouting combine" among other black scouts when he worked for Minnesota, sharing information and saving mileage if players weren't worth seeing.[37] Most scouts may not have talked much about their physical bodies, but that didn't mean they were irrelevant.

――――

If scouts didn't talk much about their own bodies, they developed colorful jargon to talk about prospects' bodies. They called a prospect a "batboy" to suggest a scrawny physique or "cakey" or "doughey" if he lacked muscle definition; such players might have "cankles" or "heavy legs." "Dead body" indicated slow reflexes, while "wound-tightness" evoked the potential energy of high-strung racehorses. Scouts could categorize bodies more formally, too. Tony Lucadello explained that scouting required properly "analyzing" the body, dividing it into eight sections—four quadrants (left and right, upper and lower) for both the front and back—each of which was scrutinized.[38]

Conventions of written reports required scouts to capture a physical description in just a few words. Evocative phrases called out certain elements for praise or criticism. "Tall Rangey Long Armed, Has Big Hands and Long Fingers." "Big Strong Well Proportioned." "Long arms, big hands." "Compact stocky body with solid lower half. Width in thighs and butt. Still somewhat fleshy looking. Has some strength in arm and should fill out well in upper body." These descriptions were distinct from the listing of height and weight; they were more about characterizing the physical

form as that of an "athlete": "Good athletic Body." "Has the carriage and actions of an athlete." "Good Agile Body." "Live body!
Extremely Loose." "Some softness." "Strong; Rugged."[39] These
terms functioned as precise bodily descriptions and constituted
an argot shared by scouts and their superiors in the front office.

These descriptions also often called upon proverbial wisdom
concerning the worth of certain types of bodies. The Washington
Senators, for example, told scouts to follow the rule that a "good
big man is always better than a good little man." Other clubs suggested that "little righties aren't prospects"—"little" here being 5
feet 10 inches and 170 pounds. Scouts used the term "good face"
as shorthand for desirability, the physical face serving as a synecdoche for physical, mental, and moral traits. Gary Nickels described the "good face" as meaning "that he looked athletic, like
a high stage in evolution—that he struck you right away as strong,
forceful, manly, open instead of withdrawn. You might see that
in his whole bearing, the way his body kind of opened out. But
you might see it in his face, too: the broad features, the strong
jaw, the direct eye-contact." Julian Mock was even more specific
about what he evaluated: "His eyes. I always judge the eyes. For
example, you watch a guy go into the dugout and he isn't even
paying attention to the game. Shakespeare said in Macbeth that
you can see the structure of the man in the face."[40]

Physiognomic types functioned as signs of value. Although the
contours of "athletic" were never precisely defined, its referent was
always current professional baseball players. Scouts could make
this comparison explicit by noting that a prospect "carries himself
like a pro." They might even compare his body to that of a well-
known professional, implying that the prospect might eventually
produce the same kind of results: "Similar to Braves' Woodward:
Tall-Trim." "Similar to Bobby Grich Type Body." "Something of a
Todd Helton look." "Reminds of Kenny Lofton."[41] Scouts' sometimes colorful language for describing bodies was in fact finely
tuned to efficiently communicate whether or not a prospect had
a physical form consistent with that of a professional ballplayer.

For amateur scouts, the emphasis was on the future prospects of the body, not its current performance. Scouts were not looking for the best amateur players (for which performance statistics would be fairly useful), but for those who might develop into professional players. A player who has peaked in high school or college will almost certainly never play professional baseball, no matter how good he looks. The trick was figuring out who had hit his peak and who was still developing. Derek Ladnier, when scouting in the 1990s for Atlanta, recalled this distinction when he watched a struggling Kevin Millwood pitch: "I mean you look at his numbers and go 'Well, this guy's not a prospect.' . . . And then when you look at him on the field and you watch him pitch and you go 'Yeah, this guy's a prospect, and a big-time prospect.' He was just learning how to pitch."[42] A mediocre player who had time to grow and develop—and the right physique—might yet become a professional athlete.

What clubs were hoping for, in any case, was the impossible. As Ellis Clary lamented, "Today everybody's looking for a nineteen-year-old boy with twenty years['] experience." Experience and development time often worked against one another: the more information a club had on how a player would develop, the less time coaches had to actually develop the player before he got too old. One concrete consequence of the need to project future performance was that many clubs chose to sign older players. Indeed, by 1977 more college players were being drafted than high school players, and by 1981, the ratio had become nearly five to one. Some of this was because the quality of college play had improved and college players were relatively cheap, but it was also because clubs desired more accurate information about a player's trajectory over time.[43]

Figuring out future prospects—whether good habits could be formed, discipline inculcated, muscles grown, instincts honed, and health maintained—was a gamble. Gib Bodet, a longtime scout for the Dodgers, described scouts' reports as similar to "the morning odds at the race track."[44] Scouts were responsible for setting odds—for accurately valuing players—but had little control over which players would ultimately succeed at the professional level.

As Guy Griffith and Michael Oakeshott pointed out nearly a century ago in their *New Guide to the Derby*, however, that's not to say that there was *no* actionable intelligence to go on when setting racetrack odds. (This tongue-in-cheek guide was meant to replace the "Reign of Chance" with an "Age of Reason" in horse betting.) Even though a horse hadn't participated in that race, with that field, in those circumstances before, a prospective gambler did know the horse's performance in previous events, its breeding, and its physical form. Races like the Derby were not typically won by horses that had never competed or had only performed poorly in the past. A prospect wouldn't have been a prospect unless he had already proven himself against the best available competition. The goal was to project future performance on the basis of the past. "Reasonable" selection of horses, like the selection of prospects in baseball, entailed knowledge not solely of past performance but also of form: whether a particular physique was well suited to future success.[45]

Underlying Griffith and Oakeshott's book was the sentiment that though these heuristics for predicting success were hardly foolproof, they had more validity than random guessing or superstition. The evaluation of prospects, like that of horses, involved heuristics honed by watching many prospects develop over time. Skillful scouts, like racetrack wagerers, needed to know what information was relevant for predicting future success. One had to be able to read the signs.

As with picking horses, coming from good stock—or resembling good stock—mattered in baseball. Some clubs, like the Atlanta Braves in the 1990s, explicitly preferred scouting children of major leaguers. There were at least two benefits to this approach, Atlanta scouts like Paul Snyder believed: first, it would give the team a clue to how the prospect might grow over time, and second, these prospects would already have a sense of the job and would be better prepared to face the challenges of professional life. There is, of course, something to the idea that children of strong, tall, and athletic parents may become strong, tall, and athletic themselves. That is, in the parlance of both scouting and horseracing, certain competitors simply had good "bloodlines" and "pedigree."[46]

Other prospects had characteristics that were deemed problematic or less desirable. Scouts had to recognize these warning signs and report them accordingly. Some scouting reports asked scouts to list problems within the category of "Physical Description." Other forms, like the one used by Baltimore in the 1970s, separated out "Injuries and Impairments" from "Physical Description" and included places to indicate "Glasses," "Contacts," and "Extent of Eye Deficiency." Tony Lucadello claimed that he wouldn't recommend prospects if they had glasses because he thought that meant they couldn't hit certain pitches, though his stance eased after he saw Chris Sabo's success with goggles in the late 1980s.[47]

Similarly, a history of injuries was seen as a sign of possible frailty, so scouts would report whether a player had any known injuries or "physical defects." Bodies were described as more or less prone to future injuries in notes such as, "Strong durable type that can withstand daily grind." Scouts had to make a judgment call based on their reading of the body and its previous injuries as to whether players might have future problems. Often this required an analysis of previous injuries, combined with an assessment of the player's present body type. It was this kind of thinking that made Boston's Joe Stephenson recommend the club "pass" on Frank Pastore in 1975 because of the "risk" of "possible elbow chips," made more likely because the pitcher's arm motion still involved "extreme whiplash" and "elbow strain."[48]

Other signs were interpreted negatively but were perhaps not as obviously problematic as injuries or physical impairments. Many reports included a box for "married," a negative attribute because a married man might demand more money for his family. There was already a claim on his labor, in a sense. Similarly, a too-old body was generally rejected. Different teams had different standards for what was "too old," though generally once an amateur player had reached his mid-20s, clubs wouldn't bother with trying to draft him because there wouldn't be enough time for him to develop before his body's natural deterioration began.[49]

Some attributes of bodies immediately made them ineligible to be scouted, however skilled they might be. For much of the century, "black" bodies were deemed problematic. Of course, there was the initial problem of explicit segregation. As scout Hugh Alexander explained, "When I began scouting, there was no such thing as signing a black player. It was taboo, if you want to use that word. It was simply against the unwritten and unspoken rules in major league baseball." Though the introduction of black prospects into the sport might have potentially reduced labor costs for clubs, lingering racism ensured that it was a very gradual transition.[50]

Even after the reintegration of the major leagues, scouts carefully noted when bodies were something other than white. A report on future superstar Bob Gibson from 1956 (when he played for Creighton University) made sure no one missed the fact he wasn't a white player: his name was given as "Robert Gibson (Colored Boy)," and his nationality as "Colored Boy." It was once commonplace for African American players to have their "nationality" noted as "colored," though later reports changed the entry to ask directly for "race."[51] For years after this change, however, only black or Latino players had their race indicated, whereas scouts often left white players' race blank, as it was not a sign that affected the player's valuation.

Scouts were required to deftly read bodies for indications of other factors that might affect prospects' value. Sometimes these "warning" signs would be presented so matter-of-factly as to be jarring to modern sensibilities. Philadelphia Phillies' reports from the late 1960s simply had a section with instructions to "check" the relevant box if the prospect suffered from the conditions of "married, glasses, or negro." The form's brevity was indicative of the reality that scouts had to look at many prospects quickly and therefore had very limited time to evaluate each one. As Leon Hamilton, after more than a half-century scouting, reflected in the early 1980s:

> In the old days you'd go into a town till you either won a guy or lost him. . . . Nowadays scouts are a little bit like the ticket sellers at the racetrack, where it's more of a computer deal. You have to do broader coverage, travel even more than before, so you don't see as much. You don't know if a boy smokes pot because you're not around town long enough to find out. You can't tell if he has heart. You don't even get to know his mother.[52]

Gib Bodet had made a strikingly similar analogy when he compared scouts to the handicappers at the horse racetrack—the ones who established the morning odds. Scouts were still playing the racetrack game, but they weren't making the decision to buy or sell horses themselves. They were just quickly handicapping the field, assigning odds, and selling tickets to the show.

———

In the 1980s Leon Hamilton regretted, above all, not having the time to *see* as much as he had in the "old days." The term "scout" originated in a military context, derived from the old French escouter, "to listen." Generals sent scouts forward to listen for information about the enemy. By the time the term was applied to baseball, however, it was the eye, not the ear, that dominated metaphors. The best scouts were defined by superior vision, by being able to see what others would miss. Vision is central to scouting—it's implicit in the "ivory hunter" and "diamond in the rough" metaphors. Tony Lucadello used to explain his habit of searching for coins on the ground before a game: "When I can't find any more pennies," he claimed, "that's when I'll know I don't have the eyes to be a scout. That's when I'll hang it up."[53]

More commonly, vision talk among scouts was about how to watch prospects themselves. Lou Gorman, a longtime scout and general manager, noted that from the perspective of the front office, the time scouts had to watch prospects was essential. "If I go out and look at a player for two games and I've got a scout that's seen him for 10 or 15 games and he likes him, I'd better stick with his judgment," he declared. Fred McAlister agreed that time watching prospects was essential: "The only way to know about a player is to see him, the way he runs, his coordination, and all his baseball skills." Even, perhaps especially, in cases involving a phone call about a player, scout Cecil Espy explained that "I don't ever do my scouting over the phone like that. I always want to see the kid for myself." Veteran Cubs scout Lennie Merullo emphasized that "[you] have to scout with your eyes and not with your ears"—that is, a scout should doubt what he heard until he saw it for himself.[54]

Not all views were equal. Buck O'Neil claimed that a scout might learn more about a player by watching him in infield practice or batting practice than by seeing him in a game. Some situations were easier than others to witness—it might take many games to observe players in high-pressure situations or batting against hard-throwing left-handers. Jack Doyle, who scouted for the Chicago Cubs during the 1920s and 1930s, when night games were becoming more common, claimed that at least in one case, he scouted a player over five games to be sure that he could translate what he saw "under the arc lights" into day games at Wrigley Field. Similarly, Lucadello claimed that things look quicker, faster, and better under the lights—but such visions are "illusions." Brandy Davis wanted to see pitchers in day games since only with natural light could he "really tell what his pitches do."[55]

Scouts and clubs had systems for viewing players. Lucadello, who previously pointed out the eight sections of the body to watch, also claimed that there were eight positions on the field to properly view those sections, with three down each foul line and two beyond the outfield fence. "You'd be surprised," he claimed, "at the difference you see in a ballplayer from one angle to another. It can affect your whole perspective." Only after having seen a ballplayer from "all sides" could he determine whether a weakness might be correctable over time; such a determination Lucadello "could never make by sitting in the stands." Julian Mock thought of his own practice of viewing prospects in terms of triangles:

> I want to be in a position with the wide-angle lens of my eyes through the hitter straight into left field, taking in the short-stop and the left fielder. That's one triangle. Another triangle is the shortstop, second baseman, and center fielder. . . . When you pick out the ones you're interested in, I want to see him everytime he does something on the field, and in the dugout, too, if I can.[56]

Whatever system they used, many teams asked scouts to characterize bodies in terms of visible "tools." The Cubs' manual from the early 1990s explained: "Five major areas are running, throwing, hitting, power, defense. It's a rare player that owns all five. . . . A major league prospect should own at least three of the tools." The club specified not only that each report must have a complete "word description" of the player, but also that each player was to be credited with his "proper tools." The Phillies' 1978 manual described the "tools" as parts of the body: "Theoretically every scout should recommend only the prospects who have the following four basics necessary to become a Major League player: The arm, the legs, the bat, and the hands." The fifth was the heart—the "great intangible" of "desire."[57]

Unlike the earliest scouting manuals, which assumed that scouts would be able to sign players on their own, later manuals made plain the importance of properly seeing prospects. The Rangers' 1990 scouting manual, for example, even explained where scouts should sit at the ballpark: "The best starting place for getting a complete view of both teams is a seat behind home plate, about half way back from the grandstand front row." From this position, scouts could see the pitcher's arm action, as well as whether the batter stepped back in fear from pitches or hit on his heels. Then, if the scout was observing a pitcher, he was to move to first or third base, opposite the player's throwing hand side. If he was scouting a hitter, he was to move to the field side, facing the player.[58]

Superior vision was also required to project a prospect's body into the future, when he would be playing at full potential. The Rangers reminded scouts that "the player who goes three for four, or strikes out twelve with a little tiny curve ball isn't necessarily a prospect. Look for the player with a strong flexible body. Watch for things that he does mechanically that will eventually bring results and success." Lucadello expressed this idea slightly differently, noting that the best scouts were "projectors." A scout "can't just look at what a young man has done," he said; they "have to visualize what he can do." Tellingly, scouts' talk of "projecting" bodies again drew on visual metaphors. Scouts wasted their time looking at "performance" alone.[59]

At heart, what amateur scouts were assessing was the "normal development of a player," not the player as he was currently. Gib Bodet claimed that as a scout, "you have to look at the player and say this is what type of major-league player you think he will become. And how likely it will be that he will realize his ability." Hep Cronin said something similar: "Good scouting is being able to visualize how a boy is going to develop. That's projection. That's what scouting is all about: being able to look at the body, the ability, and the attitude of a player today and project how good he is going to be in the future as he gets bigger and older." The Phillies warned scouts that some skills were effectively fixed, however: "By the time a prospect reaches 18 years of age, his speed afoot is determined. In most cases, the same could be said for a player's arm strength, however don't eliminate the possibility of improvement, especially from a pitcher who will mature and get stronger." The club acknowledged that fielding and hitting could, however, be improved given a strong and coordinated body.[60] The premium was on seeing and reading the body well enough to visualize it in the future.

Jethro McIntyre claimed that this need for projection was why it was so important for scouts to have played baseball themselves:

> Scouting jobs are meant for guys who played the game, guys who know how to play the game, and know what it takes to be in the game. Baseball needs guys who are connected with baseball and understand what it takes so they can see the kids in that kind of light because we are supposed to predict their potential down the road and all that.[61]

This was a claim about expertise: the job of predicting or projecting required familiarity with those settings in which the player would have to perform. Scouting required the ability to see the prospect now *and* to visualize his potential in the future.

————

Scouts were bureaucrats and careful readers of the body, but they were also appraisers. The key question for scouts was, "How much is a player worth?" This information was inscribed in nearly every report any scout sent to the front office. Carew's information card indicated Stein thought he was worth $5,000. That was in some sense all that mattered—finding a number that was high enough to be acceptable to the player and low enough to be worth the investment risk for the club. One 1978 Phillies' manual put the matter succinctly when it told its scouts to fill out a "dollar evaluation" line on every report they submitted. This would "be the highest figure you would go in order to sign a player if he were on the open market. . . . Boiled down it is the 'dollar sign on the muscle' and no more."[62]

The cost of signing was ultimately a bet on a player's value: it might rise or fall. Even in the age of free agency—and certainly in the age of the reserve clause—the club controlled a player's labor for years. They could trade him, sell him, play him, or just stockpile him. The player had little or no power for the first part of his career. That's why accurate pricing was so important. Very few clubs could afford to wait for players to prove themselves in the majors and then pay top dollar to sign them. It was far more cost-effective to gain control of the labor—buy the "muscle"—of young players cheaply, on the recommendation of amateur scouts.

Though not everyone would describe it in so stark a fashion, scouts were middlemen in the buying and selling of prospects. Over the century of the reserve clause's existence, from the 1870s to the 1970s, players complained about the way it effectively turned them into property. John Montgomery Ward asked in 1887, "Is the Base-ball Player a Chattel?" His answer was a resounding "yes": players were chattel in the eyes of owners and could therefore be "disposed of as a valuable commodity" within a system that openly put them up "for sale."[63]

Nearly a century later, just after another major expansion of civil rights, outfielder (and African American) Curt Flood echoed Ward's language in rejecting a trade from St. Louis to Philadelphia. As he wrote to Commissioner Bowie Kuhn in 1969, "I do not feel that I am a piece of property to be bought and sold irrespective of my wishes."[64] In the midst of his own contract dispute, Flood found that the restrictions on his ability to sell his own labor reduced him to a piece of "property."

It was the scout's duty to price this property and—at least before 1965—often to make the purchase. Hugh Wise's scouting report for Gary Bell in 1955 noted that Bell was currently "Cleveland Property" but "recommend[ed] purchase or trade in case opportunity presents itself." In these circumstances, prospects "performed commodification" in a manner analogous to historian Walter Johnson's description of slaves' behavior at slave markets. Even in horrific circumstances, enslaved persons had the ability to display their body—its attributes, virtues, and vices—in such a way as to help them navigate among bad options.[65] Prospective athletes, although obviously far more privileged and autonomous, were similarly aware that scouts were watching them in order to price their body and its labor.

Prospects indeed may have played differently, spoken differently, or displayed their body differently knowing that scouts were around. As Buck O'Neil claimed, "Believe me, once the players knew there was a scout in the stands, they performed differently." This was, in fact, exactly the advice players were given in manuals like *How to Make Pro Baseball Scouts Notice You*, which informed readers about how to "market themselves" and what to do to "impress" scouts or "what not to show the scouts." There is a long history of humans self-tracking in an attempt to create desirable outcomes through the production of data about themselves, from public scales to Fitbits.[66] The data produced about prospects were no different: prospects knew their actions would be turned into marks on a page—and eventually into a price—and displayed themselves and their skills correspondingly.

Scouts undoubtedly recognized that signing a player was not the same as buying an automobile or a lamp, but the end goal was similar in that one needed to find a desirable product at a good price. Indeed, the ability to price players accurately was a key factor separating good scouts from bad. As Frank McGowan explained in a 1960 talk to fellow Baltimore scouts entitled "Judgment—The First Ingredient in Scouting," a scout's own value was largely determined by his ability to properly evaluate players: "Overvaluation is a waste of money which might buy another player. Undervaluation robs the club of a valuable piece of property."[67] Scouts may have believed themselves to be in the talent business, but clubs often considered them to be in the valuation business. Recognizing talent but failing to price it accurately rendered scouts' judgment all but useless.

The reports scouts submitted to the front office made this role explicit through their inclusion of an "Evaluation" line. The most common forms used after the 1970s had lines for "Worth" as well as "Will Sign For," enabling a scout to distinguish what a player was worth from the lower amount that he would (hopefully) be willing to take. Kansas City reports from the 1960s cut to the chase and simply asked scouts to specify "What We Should Offer" players. Houston's reports from the 1970s likewise had a space for indicating "What's He Worth?" Other teams used a "Money Expected" line. Milwaukee's reports from the 1990s used a series of check boxes for "Value" ranging from "$200,000+" down to "less than $5000"; Chicago White Sox forms simply had a place to put down a player's "Value."[68]

The reports prepared on Biggio fell into this pattern. The Cubs' scout, Blitzer, noted Biggio's "dollar value" right at the top of his report on him—$65,000. Another scout, Brad Kohler, believed he was worth $85,000. The Angels' Jon Neiderer thought he was worth only $60,000 and too costly even at that price. In the end, Biggio was signed by the Houston Astros in 1987 for $110,000, much less money than was paid for many of the 21 players selected before him, including Mark Merchant, signed for $165,000 by Pittsburgh, and Chris Meyers and Dan Opperman, who each signed for $160,000 (with Baltimore and Los Angeles, respectively). None of these prospects ever played a day in the major leagues.[69] Even though scouts agreed on the fact that Biggio had the talent to play, the disagreement on his value made all the difference (though the Astros would still end up paying him over $80 million in salary over the course of his career). Scouts were jewelry appraisers, not diamond hunters.

———

Economic theory seems inappropriate for the pricing of prospects: clubs don't appear to be rational actors expressing preferences about known quantities. Indeed, as the economist Lucien Karpik has shown, that's in part because objects like athletes' bodies aren't like televisions or toasters. They're "singularities"— unique and largely incommensurable products that don't fit into any traditional economic model. Scouts' work to price prospects is more like the process of pricing works of art and fine wines than that of pricing mass-produced widgets. The effort to price "singularities" requires devices of judgment and trust, since the "true" value of prospects is in a fundamental sense unknown and unknowable. Some will undoubtedly be worth millions of dollars after a few years, while most will be effectively worthless.[70] This is one of the factors that most sharply distinguishes amateur from professional scouting: When evaluating a player who has experience in the major leagues, performance statistics are often useful. When considering prospects who have years to develop and who have never played against comparable talent, amateur scouts must find ways to quantify other factors.

Scouts know this without needing to know any formal economics. They know that they are attempting to put a price on a body based on necessarily incomplete, often sketchy information. Throughout the twentieth century and into the twenty-first, they have needed to properly price players regardless of whether they have been restricted by a draft or only by the depths of an owner's pockets. They also know that there can be no hard and fast rules because each case is different in crucial ways.

Paper technology itself plays a key role in such complicated decision-making, acting as an information management technique, a way of organizing how scouts see and value prospects. Though scouts emphasize the visual aspect of their work at the expense of the written, there is a close historical link between observation and the written word. Taking notes for scouts, no differently than for natural philosophers or doctors, means fixing one's view of the object, cropping out what's unimportant, and emphasizing relevant details in such a way that they can be compiled and reread in juxtaposition with the observations of others. Note-taking and report-writing aren't separate from processes of classification or evaluation: they are essential to them.[71] Scouting is a practice of using recordkeeping, trained observation, and accurate evaluation in order to price prospects.

Scouts' public image as loners on dirt roads searching for hidden talent is not unlike age-worn tropes about scientists as solitary inventors or mad geniuses. The typical reality of each—carefully monitored labor practices with extensive paper trails conducted in the midst of corporate bureaucracies—is far more mundane.[72] That's not to knock either science or scouting. That's just how reliable knowledge usually gets made.

6

Measuring Head and Heart

Nearly every report filed on Craig Biggio commented positively on his "makeup." Habits, aptitude, maturity—these characteristics provided crucial justification for drafting Biggio. By the time of his junior year at Seton Hall, he was known as a dedicated, smart, and mature player. He was, in short, a better draft pick than another prospect with the same statistics.

One reason scouting knowledge seems so different than statistical knowledge is the persistent belief that scouts can gain access to something no sabermetrician ever could: a player's psychology. When longtime scout and general manager Lou Gorman read Michael Lewis's *Moneyball*, he particularly resented that Oakland general manager Billy Beane had "flirted" with the idea of firing his scouting staff and drafting players using a computer. That was foolhardy, Gorman thought. Pointing to his head and then to his heart, he said, "Well, you can't measure what's here . . . on a computer. There's no way in the world you can do that." Some scouts described this skill as the ability to recognize the "good face."[1] By this they meant that they could see something about a player's psychology or physiology that could not or would not be visible in statistical records of on-field performance. Statistics alone would never reveal what a scout ultimately wanted to know—whether a player would succeed at the highest levels.

Some bodies, after all, harbor secrets, sometimes harmful ones: hidden injuries that will shorten a promising career, physical tendencies that will put a ceiling on future development, mental weaknesses that will hamper any success. Others hide helpful ones: the ability to be coached, the possibility of physical growth, the mental strength that can transform a mediocre infielder into a team leader.

Whether evaluating the "good face," charisma, character, or drive, scouts are supposed to specialize in uncovering the signs that will enable accurate evaluation. The crux of Lewis's book is Beane's attempt at redemption from his own fate as a highly touted, highly skilled player who ultimately failed because he couldn't handle the pressure. Though there were signs that his performance was dropping off before he was drafted, the statistics alone couldn't have indicated whether Beane would succeed at the top level. Art Stewart, a longtime Kansas City scout, acknowledged that "numbers are nice, and you need to know the numbers... but you can't base your evaluation on those numbers, especially when you're talking about high school kids."[2] Every high school prospect has absurdly good stats. The problem is determining which of them have the right psychological and physical attributes to perform at the professional level. This was always the underlying irony of Beane's epiphany about the value of dispassionate statistical analysis.

Sabermetrician Bill James has made a related point about prediction in a number of different ways over the years. In 2013 he described it as a fundamental gap in statistical models. Set the performance level of major leaguers as 1,000, he explained. If college athletes played at an 800 level, it wouldn't be difficult to predict performance: it would simply be a matter of scaling and projecting. But it would be more accurate to say that college competition is really below 50 on that scale, and high school competition even lower. Statistics are simply not useful in this situation; projections are too distorted. Deciding which statistical records will translate into success at the professional level requires more than knowledge of the statistics themselves. That's where scouts come in.[3]

Though it may be tempting to conclude that scouts focus only on the subjective aspects of player evaluation, clubs and scouts have sought for decades to find objective, often mechanical means of revealing these otherwise hidden physical and mental signs. With this goal in mind, they have long looked for psycho-motor tools that might cheaply and reliably assess such traits and have called upon psychologists, physiologists, and other scientists to measure "intangibles." Scientific—that is, technical—knowledge has long held promise for making judgments about players' physical and mental "makeup." The invention of radar guns, the introduction of psychological testing, and the establishment of research programs for baseball all promised that with the right tool, clubs might more consistently, and more cheaply, identify prospects.

Ultimately, however, clubs failed to replace the expertise of scouts with mechanical tools, training academies, or standardized tests. Rather, these new innovations themselves required great skill to operate and interpret. They became yet another way scouts and clubs attempted to turn judgments about prospects into reliable numbers.

———

Many scouts have described visible and invisible attributes as two sides of the same coin or, as a now-famous chart from the 1960s would suggest, two halves of the whole. Baltimore scouting director Jim McLaughlin designed his "Whole Ball Player" chart to emphasize that scouting requires an analysis of both what can and what cannot be seen with the eye. The usual measures of a pitcher's fastball and control are visible, as are an outfielder's range, speed, and power and all players' hustle, size, poise, and coordination. But some things can't be seen, particularly attitude, mentalities, personalities, and background.[4] Scouts want to know a player's psychological makeup but cannot often witness these qualities with their own eyes.

Some scouts swore by the chart. Art Stewart claimed to have it memorized, with a (apparently unnecessary) copy framed at home, because it was a "good way to picture" what he could see with his eyes versus what he had to "feel" and learn over time. Lou Gorman similarly said that he had used the diagram during his days in Baltimore and claimed that it combined ideas from Baltimore, Kansas City, Los Angeles, Seattle, and other clubs' scouts. "It was a way to show the scouts about the nonphysical skills," he declared. Though not citing the chart explicitly, when scouts like Ray Scarborough referred to the "lower half," they were talking about this same idea, the "personal makeup of the individual player—what he's like inside, how he lives, what he believes in."[5]

Others rejected the idea that makeup was invisible, claiming they could reveal it with the right stratagem. Scout Al LaMacchia explained that when he was testing a prospect's "makeup," he would invent a story about the prospect walking away from a confrontation. Then he would tell the prospect's teammates the story to see whether they found it plausible. If they didn't, LaMacchia interpreted it as a sign that the prospect had competitive drive. Clubs that aspired to look specifically for "makeup" in players, like the Atlanta Braves in the late 1980s and 1990s, also believed that it could be seen directly. The pitching coach for the AAA affiliate in 1986 claimed that when future Hall of Famer Tom Glavine played for them, "you could just see his makeup. . . . You could tell he was something special." And when General Manager Bobby Cox was choosing between Chipper Jones and Todd Van Poppel as the first pick of the 1990 draft, he claimed he chose Jones after the prospect decked an opposing player who dared to yell something at Jones's teammate. "Well, the makeup question about whether or not he's tough enough has just been answered," concluded a Braves scout afterwards.[6] An unwillingness to back away from a fight, even in a nonbaseball setting, was a sign that a prospect had the "makeup" to be a professional.

McLaughlin had his own ways of making visible this "lower half" as part of the "Oriole Way" of scouting the "whole" ballplayer. Whether deploying psychological tests, bringing in FBI consultants on how to conduct background searches, or sending managers to Dale Carnegie–inspired seminars, he was committed to getting people on the team to think about the mental side of the game. McLaughlin saw scouts as "damn unscientific" and instead "wanted to inculcate basic principles." "I wanted rationality," he said. "I wanted science."[7] For McLaughlin, rationalizing scouts meant finding ways to make physical skills and intangible qualities visible and measurable.

The idea that science itself might provide a way of selecting players and predicting success goes back a long way in baseball, certainly to Henry Chadwick's description of the sport as a manly, scientific game, but it became widely emphasized in the twentieth century. Books with titles emphasizing the "scientific" side of baseball—such as Byrd Douglas's *The Science of Baseball* (1922), Ted Williams's *The Science of Hitting* (1970), and Mike Stadler's *The Psychology of Baseball* (2007)—claimed to offer a unique or particularly incisive look at the game.[8] Sometimes these volumes drew from actual scientific findings, but most commonly they treated "science" as "systematic knowledge" or rationality writ large. Often written as "myth busters," these books claimed that baseball had a built-in advantage for those who approached it systematically.

As early as 1910, baseball's *éminence grise*, A. G. Spalding, was interviewed by the *New York Times* on the connection between baseball and experimental psychology. Citing new college-level courses on psychology, he described the field as exploring "the relation between thought and action [as] recorded by delicate instruments." Because Spalding believed baseball was both a mental and a physical activity, the new techniques and technologies of psychophysiological measurement developed in the psychological lab offered the opportunity to potentially measure which prospects had the most promising reactions.[9]

The same year, the sportswriter Hugh Fullerton was also effusive in his praise of the use of science to understand the game. In his book *Touching Second: The Science of Baseball*, co-written with Hall of Fame second baseman Johnny Evers, Fullerton expressed his belief in the potential of science to fundamentally improve both the way the game was played and the way prospects were identified. In it he told the story of Tom (Toad) Ramsey, who was "discovered scientifically" in the 1880s when the manager for Louisville's team, James A. Hart, noticed that Ramsey could twist the cover off of a baseball with his fingers. Such strength led Hart to look for prospects among other "left-handed brick-layers" rather than among minor leaguers, although he failed to find any more.[10]

Fullerton and Evers also claimed that an understanding of psychology was needed for success in professional baseball. They quipped that "Professor Münsterberg," the famous director of the Harvard Psychological Laboratory brought stateside from Germany in 1892, "could save managers and club owners much time, trouble and money, as well as many disappointments, by testing the brain action of players psychologically and discovering their brain speed before a season opens." The psychologist's promise to reveal the "hidden" workings of the mind had considerable appeal given their contention that "baseball is almost as much psychological as athletic." Though showing no real understanding of contemporary psychology themselves, the authors believed its findings might be profitably incorporated into the game. This view was in no small part due to the advocacy of Hugo Münsterberg himself, who consistently promoted the many services academic psychologists could offer to corporations, municipalities, and, of course, athletes.[11]

Fullerton would go on to be an enthusiast for science and scientific approaches throughout his career, especially in his breathless coverage of Babe Ruth's trip to a psycho-motor laboratory in 1921. Fullerton's cover story in that October's *Popular Science Monthly* promised to finally answer the question of "Why Babe Ruth Is [the] Greatest Home-Run Hitter." After an afternoon game, Ruth had been subjected to a battery of then-standard psychological tests by Albert Johanson and Joseph Holmes, two psychologists in a Columbia University laboratory once run by James McKeen Cattell. Trained by the famous Wilhelm Wundt in Leipzig, Cattell was the first to hold the title of professor of psychology in an American university and a central figure in the spread of psychology across the country, particularly through his experimental work on response times.[12] Though Cattell had departed by the time of Ruth's arrival, the lab stood for the highest level of psychological expertise.

Fullerton's account of the experiments explained how Ruth scored consistently well on tests of response time, perception, and coordination—"every test but another triumph," he concluded. Despite the lack of evidence concerning how well Ruth actually performed during an exhausting three-hour interrogation (most of the time he was only average, and even when differences were noted it was not clear which, if any, were meaningful), Fullerton was convinced that the ballplayer's psycho-motor response could be shown to underlie his ability. He remarked that Ruth might even use the results to improve further—apparently, Ruth held his breath when hitting, meaning he wasn't using "maximum force."[13] Otherwise, Ruth showed himself able to respond rapidly to a flashing light, jab a tool into holes on a board, and decipher a string of eight letters shown for only one fifty-thousandth of a second. These tests were all considered standard by the 1920s, used to measure telephone operators, typists, and telegraphers, and the *Popular Science Monthly* story, along with coverage in the *New York Times* and elsewhere, only confirmed the role of scientific analysis in understanding what makes certain people excel at difficult tasks.

Fullerton labeled the experimenters "scientific 'ivory hunters,'" making the link to scouting explicit and implying that perceptual quickness was indeed a key part of Ruth's "secret" to success. He suggested that club owners might "take the hint from the Ruth experiments" and "organize a clinic, submit candidates to the comprehensive tests undergone by Ruth, and discover whether or not other Ruths exist." Though he did not use the term directly, Fullerton was writing about psychology's claim to be able to measure "aptitude" for various activities, determining which people were best suited to become stenographers, engineers, or infantrymen.[14] For baseball clubs, the sciences—particularly the psychological and physiological study of perception and coordination— might be put in service of finding future stars.

Fullerton's piece had the potential to inspire a generation of researchers traveling around the country looking for baseball players with lightning quick reflexes. He was clearly pushing for the development of laboratory-based shortcuts, quick tests that might uncover diamonds in the rough. Psychology's methods offered a great opportunity for baseball's managers and owners, Fullerton claimed: through psychological testing, they could quickly and scientifically identify those with the most aptitude for the game, as well as those better suited to another line of employment. Indeed, such tests were proving their worth in other fields. Psychologists were beginning to exert a dominant influence in education, and Columbia—where Ruth underwent his tests—would soon become a central node in the development of aptitude tests, including what would eventually become the SAT. Four years after Fullerton's article appeared, psychologist Coleman Griffith predicted that it was only a matter of time before a similar test was able to select the "best adapted" men for athletic excellence.[15]

Few others followed Fullerton's lead, however. Only a handful of academic and quasi-professional psychologists between the 1920s and the 1950s advised clubs, the most famous being David Tracy and Griffith himself. Griffith, unlike Tracy, had a sterling "pedigree," with mentors who also claimed a connection to Wundt's laboratory in Leipzig. Both Griffith and Tracy were hired to advise clubs (the Chicago Cubs in the 1930s and the St. Louis Browns in 1950, respectively) and wrote books detailing their advice (*Psychology and Athletics* and *The Psychologist at Bat*, respectively). In both cases, their emphasis was less on finding future stars and more on encouraging behavior modification through training in mental and perceptual skills. As Griffith explained, "The athlete who goes into a contest is a mind-body organism and not merely a physiological machine."[16] Both claimed their services offered financial savings, in that it was cheaper to hire one psychologist able to encourage peak performance in players than to waste money on expensive yet underperforming stars. Neither Griffith nor Tracy, nor anyone else before the 1960s, however, did much to actually establish a sustained psychological study of baseball.[17] Neither was successful in his attempts to change the performance of even one team; both blamed their failures on the intransigence of baseball personnel, particularly managers.

Whatever these psychologists' failures, by the 1960s some clubs thought it was obvious that they needed to invest in science in order to streamline the process of scouting thousands of players, drafting dozens, and then hoping that one or two would ever play a day in the major leagues. Bob Carpenter, then-owner of the Philadelphia Phillies, paid for a team of University of Delaware professors to conduct a "Research Program in Baseball." In introducing the project in the first status report of July 1962, the principal investigator (and professor of physical education and athletics), Harold Raymond, wrote that "there is evidence to suggest that a program of scientific investigation could make an important contribution to an objective measurement program which could, in turn, significantly improve the chances for success in selecting potential major league players." Working with Phillies personnel, Raymond and his colleagues (particularly Robert Hannah, then a research fellow in physical education and athletics) attempted to find "which factors might logically be related to success in baseball."[18]

Raymond and Hannah were not looking for just any factors, however—they were looking for "objective" factors. They noted that scouts did use some "objective" factors, like batting average, but that all too often, the scouts' final measure of players was a "subjective and personal one which is susceptible to a variety of rater errors." The project, then, was to find those measures—psychological, biographical, physiological—that might enable a team to know which players were most likely to be successful without succumbing to scouts' claims of special expertise. It was difficult, after all, to "evaluate another person objectively without having that opinion distorted by one's own image." Though the researchers were doubtful they could find a "panacea" to this problem, let alone entirely replace "trained observers," they did hope to find more objective data from which "superior decisions" could be made.[19]

The Delaware team looked to create objective data through the tools of the psycho-motor trade. They self-consciously presented their work as following in the tradition of psychological experimentation, quoting admiringly the pioneering psychologist E. L. Thorndike's statement that "whatever exists at all exists in some amount."[20] Consulting with Raymond was also Paul Smith, chief of a physics division of the nearby DuPont laboratories, and the group was clearly meant to represent academic-scientific expertise. In its first report, the team decided to explore four devices: an "ortho-rater," which could quickly measure visual acuity and depth perception; a biographical inventory of minor league players, which could detect whether certain socioeconomic or psychological factors correlated with success; a recording of a controlled hitting situation, which could measure what constituted successful swings; and a head-mounted camera, which could track the way a player's eyes followed the ball. Initially focused on University of Delaware players, the hope was to eventually recruit professionals.

Suffice it to say, the studies didn't go precisely as planned. The ortho-rater experiment demanded "further attention," while the head-mounted camera, which required constant expert adjustment and calibration, was almost immediately shelved. Researchers also soon concluded that the controlled hitting situation was far more difficult than they had realized. The accelerometers failed as soon as the bat contacted the ball, and cables attached to the bats appeared to inhibit the hitting action altogether. A radio transmitter was then developed to relay impulses from the bat's movement, but at the cost of a player having to use a hollowed-out bat with the transmitter inside. The unit then suffered two cracked resisters and several broken wires after only 25 pitches.[21]

Perhaps most successful was the biographical inventory. Nearly 150 responses were collected from minor league players, enabling the researchers to start finding correlations with player success over time by recording information on punched cards that could be analyzed on a computer. The results, though robust, were less than groundbreaking: they discovered that most minor league players were between 19 and 21 years old, were taller than 5 feet 10 inches, and had started playing baseball by age 11. Researchers continued to break down data to figure out whether some biographical aspect or another might be predictive.[22]

Though most of the research amounted to very little, after five years the Raymond team could claim to have learned something. Superior hitters seemed to wait longer before striding, put their stride foot down sooner, and have a faster swing—although the differences were not statistically significant and could account for only a small fraction of actual batting performance. The orthorater did prove to be useful in directing players to undergo full eye exams, and there was at least anecdotal evidence that among players who were released by the club, a relatively high percentage of them had also failed an eye exam, though this finding was hardly predictive. The biographical study continued to garner lots of data, though it was of questionable use to know that an average professional player was 26 years old and 191 pounds, had a parent who made between $3,000 and $6,000 annually, and was most likely to name his father as the "person most responsible for early development." The program was subsequently quietly disbanded, and when Kevin Kerrane asked about it in the early 1980s as part of his research for his history of scouting, the Phillies' assistant scouting director didn't know or care what had come of it. Despite its university imprimatur and expensive tools, it simply hadn't provided evidence that laboratory technologies could improve upon the way scouts evaluated players.[23]

A decade after the Delaware program, Kansas City Royals owner Ewing Kauffman created a baseball academy as a solution to the expense and risks of signing top prospects. There were certainly plenty of excellent athletes who simply didn't play baseball, Kauffman reasoned, and if it were possible to identify them and train them to play the game, this might reduce player procurement costs. Kauffman's presumption was that the sciences might help identify prospects hiding in plain sight—young men who had the physical and mental ability to play the game but for whatever reason simply hadn't yet. Not all prospects would succeed, of course, but any who did would possibly represent an impressive return on investment.

Kauffman, who made his money in pharmaceuticals, thought of his program as an effort to be more "definitive" and scientific about player procurement and development. The academy sought to provide both academic and baseball training, developing the prospect's mind and body. From the introduction of stretching to the use of exercises that broke baseball maneuvers into their essentials, the idea was to treat baseball as a set of skills that could be taught, based on the assumption that it was possible to identify prospects capable of performing at a professional level. The Royals screened prospects using physical and psychological tests, including 60-yard dashes, agility tests, psycho-motor tests, and basic throwing and hitting exercises.[24] Though the academy closed after a few years, having produced only a handful of major leaguers, it represented another attempt to deploy scientific testing as a screening tool that might enable teams to cheaply identify and cultivate future stars.

These initiatives were essentially meant to provide an alternative to scouts. If it were possible to isolate specific physiological and psychological factors that predicted success, then those could be used to identify potential stars, particularly ones other clubs had ignored. Not incidentally, clubs might also save money on signing players who didn't fit the traditional ideal of a prospect. Their vision of making baseball more "scientific" was a matter of finding tools for efficient sorting.

——

Though teams never widely incorporated academic psychologists into the day-to-day practice of baseball, laboratory tools, particularly ones meant to assess psycho-motor reactions, quickly became common because they promised to enable mechanized—and thus easy, efficient, and reliable—measurement of underlying ability. Tools like stopwatches had long been used as part of the "scientific management" of laborers, especially as aides for judging workers' efforts. Baseball teams turned to stopwatches for the same reason that Frederick Winslow Taylor and his disciples had: to make assessments of ability more rational.[25] Roy Clark, a scouting director for Atlanta, explained that when looking to evaluate physical skill, "we do look at stats, but it's is [sic] just a tool just like a radar gun is a tool and just like a stopwatch is a tool." As the Rangers told their scouts, there were many useful tools like radar guns that could serve as "measuring devices to aid" scouts in "making a judgment" about a pitcher's velocity. Instead of simply praising a pitcher because few hitters succeeded against him, a radar gun enabled more precise, more rapid, and possibly more objective assessment. Mechanical tools were likewise used in tryouts to efficiently sort prospects. Those who failed to run 60 yards in seven seconds or reach 85 miles per hour with a fastball could be quickly dismissed. Branch Rickey's original idea for tryouts indeed made running the 60-yard dash and throwing from the outfield to home (exactly 260 feet) on one bounce the first screening tests. Rickey had also developed a spring training facility in Florida that operated as a "factory that rolls ball players off an assembly line with the steady surge of Fords popping out of the River Rouge plant," in the words of a *New York Times* profile. Rickey and his successor, Walter O'Malley, believed mechanical tools—"gadgets" like pitching machines—could not only train players but also assist the team in selecting among them for promotion. Each team or scout might emphasize different standards and tools, but all could use mechanical devices as shortcuts.[26]

Tools still required expertise, however. They couldn't be used by just anybody. Scout Howie Haak noted that he would time players between first and third base or second and home but never between home and first, because that stretch conflated a player's pure speed with his ability to get out of the box. Scouts were instructed by the Rangers to be consistent when starting the stopwatch—"DO NOT START your watch with the batter's swing, but wait until the bat meets the ball." Mechanical tools were meant to ensure that scouts recorded physical attributes accurately, but they required instruction and discipline to yield useful numbers.[27]

After 1974, handheld radar guns provided yet another tool for making judgments of physical characteristics more precise. As recounted by Dan Gutman, the radar gun was developed practically overnight for use in baseball. In September 1974, Michigan State coach Danny Litwhiler noticed an article on a radar gun recently acquired by the local police force and asked an officer to drive over to the field and point it at his pitcher. It worked, more or less, recording the speed of the pitch as if it were a speeding car. Within six months a prototype had been developed by the JUGS company—known at that point mainly for pitching machines— and the following spring training it was placed on the standard list of baseball and scouting equipment.[28]

It wasn't that scouts couldn't quickly distinguish by sight whether a player threw hard or ran quickly. Hep Cronin, for one, didn't bother to use stopwatches, since he believed all that was important about running ability was visible to the naked eye. Other scouts and scouting directors, like Walter Shannon, simply noted that they didn't need a stopwatch to distinguish average from above-average major league speed. As Gary Nickels explained only a few years after the radar gun's invention, however, such devices still had a place:

By the end of my first scouting season I could watch a kid throw and usually come within a couple miles an hour of how fast he was. But I don't mind an accurate gun at a night game—or at a tryout game or spring training, where you see pitcher after pitcher and they all start to look alike. . . . I think radar can be a useful tool.

Similarly, the Rangers warned their scouts:

Most baseball men will agree that the running speed of a player can be deceptive. A little man taking a lot of steps might give the illusion of speed, whereas he is just a fair runner. In order to remove part of this illusion, and try to put the judgment of speed on a mechanical basis, we will want our scouts to clock all players they scout, both in professional and amateur ranks.

Longtime Chicago White Sox scout George Kachigian concurred: "Velocity sometimes is very deceptive. It might look like some guy's not throwing too hard, but then when you look (at the gun)—Holy cow!—94."[29] In this way, mechanical tools could help scouts enhance their vision, especially in imperfect conditions.

Nickels was also quick to note that many older scouts, those with clout and seniority, simply chose not to use radar guns or stopwatches as much. Instead, they went "by feel. That's partly because the younger scouts don't have as much experience to rely on, and partly because when we do have the feel we don't have the clout to make the ballclub listen to our intuitions." Ed Katalinas, back in the early 1980s, turned this statement around by noting that "ability" for experienced scouts is "naked to the eye," but some younger scouts "don't trust their own eyes, and the modern way is 'He throws 86,' or 'He has a 76 curve.' It's a number thing now, just like scoutin' is a number game now."[30] With trust established, a scouting director might simply take an experienced scout's word for a pitcher's velocity or a runner's speed, while a younger scout would have to provide evidence from mechanical tools.

There was a risk in overrelying on mechanical tools like radar guns, however. Julian Mock warned that "you can't just go on measurements. There are so many other important things. . . . And a gun is one of the worst things because some guys sit there for the whole game and gun every pitch." Instead, he recommended scouts move halfway down the first base line and sit in the first row of the stands—from there they could "see how it looks to the hitter."[31] To him, scouts saw better without the gun and, in fact, saw more of what was essential. Every other scout would be behind home plate, climbing on top of each other trying to get a glimpse of the gun. Mock simply took a better vantage point.

As Mock noted, radar guns tied scouts to one place, possibly inhibiting their ability to see the game. There was even a tradition—more prank-driven than serious—of intentional miscalibration of the gun so that anyone else naïvely looking at the gun instead of the pitcher himself would see the wrong speed, though this ploy was rarely effective. Scouts complained that so-and-so "had that thing doctored to throw us off, but he might as well *advertise* that it's six miles an hour slow."[32]

Aside from intentional shenanigans, guns were perceived as occasionally inconsistent. Nickels complained that different brands gave different results: the JUGS gun "is too skittish, but the Ray [RA-] Gun's okay." There was a technical explanation here, too, in that different radar guns measured the ball's position at different places. Different speeds naturally resulted. Into the 1990s, scouts occasionally indicated which kind of gun had been used to measure the velocity in a prospect's scouting report to ensure there was no ambiguity. The 1991 Cubs Scouting Manual clearly noted that the "RA-Gun" was 4 to 5 miles per hour slower than the "Speed Gun," though if a pitch was extremely low, the RA-Gun might read slightly higher. There have been, of course, technical improvements over time, and some scouts are now told that MLB's Statcast system will provide even more measurements of the movement of pitches, though longtime Houston scout Bob King has lamented that such "improvements" only increase scouts' dependence—"Pretty soon we'll really be sleeping with the gun." Nevertheless, as new technical systems come into use, scouts and clubs will have to figure out how to interpret the new readings; even a simple change in where a pitcher's velocity is measured can result in the need for complicated translations to compare with older readings.[33] Radar guns don't eliminate the need for careful interpretation and judgment.

However scouts used tools, their job was to cut through possible manipulation to see what was real. Scouting a pitcher meant looking for more than speed. As Gib Bodet said, "Essentially the gun doesn't tell you anything"; you need to watch with a "trained eye." Hugh Alexander elaborated that "it's still a *human* science. So I don't rely on mechanical aids, like stopwatches and radar guns. I can be accurate without 'em. . . . I want to see, period. Because to project a boy into the future, the first thing you have to do is see him in the present the way he really is."[34] A tool meant to measure current performance might actually hinder the ability to project into the future. In order to determine the value of a prospect, scouts wanted to see *him*, not just record how fast he threw a ball.

Alongside players' physical performance, teams also wanted to measure the health of the body. One clear warning sign to clubs was an injury-prone body: no amount of skill was worth the money if a player was constantly injured. By the 1970s nearly every club had created extensive forms for scouts (or, ideally, physicians) to fill out on prospects' health, from records of illness ("frequent colds?"), immunizations, and past operations to smoking and drinking habits.[35] By the start of the twenty-first century the medical evaluation had become part of formal collective bargaining, and the official "medical history questionnaire" as well as an "initial orthopedic history examination" and "orthopedic ailments checklist" were included as attachments to the basic agreement between players and clubs. The Sutton Medical History Report—also called the Athlete Health Management System—comprising more than 12 pages of detailed questions about an athlete's physical history, was also available for teams to use. New scientific research

MEASURING HEAD AND HEART 187

has enabled many players to gain strength over the years, and some stars with remarkable longevity have done much to publicize the benefits of nutrition and strength training.[36] Even as clubs have come to know more about physical conditioning, however, it remains extraordinarily difficult to predict which bodies might find success at the rarefied level of professional athletics.

———

Alongside measurements of physical performance, clubs have held out the promise of accurately and precisely measuring prospects' psychological states for decades. The formation of the North American Society for Psychology of Sport and Physical Activity in the 1960s represented the culmination of a long process of psychometric and psycho-motor tool development. As personality profile tests became common in the 1970s, psychometric testing finally began to deliver on its promise of putting a number on prospects' desires, instincts, intelligence, and discipline. The most popular test initially was the Athletic Motivation Inventory (AMI), which was widely distributed in the early 1970s by the Institute for the Study of Athletic Motivation. Within a couple of seasons of its creation, scout Joe Branzell, who worked for the Washington Senators and Texas Rangers, was carrying copies of the AMI with him as he traveled the country.[37]

The AMI had been developed by psychologists Thomas Tutko and Bruce Ogilvie at California State University, San Jose. They had noted in the 1960s that personality tests were sometimes deployed to try to learn about athletes' minds. But tests like the Edwards Personal Preference Schedule, the Minnesota Multiphasic Personality Inventory, and the Osgood Semantic Differential were not specific to athletes. The tests could reliably tell athletes from nonathletes, but, as Tutko and Ogilvie's student Leland Lyon explained in his master's thesis, "the exact nature of these differences is elusive." Moreover, "the evidence for differences between athletes—be it from sport to sport or successful from unsuccessful—is even more ambiguous."[38] Obviously coaches, scouts, and owners wanted to know which athletic youngsters were best suited to baseball and which of them were most likely to be successful. But those were precisely the kinds of questions previous tests failed to answer.

So Tutko, Ogilvie, and Lyon set out to design a personality inventory specifically for athletes, meant to measure traits that either had "empirical support" for athletic success or were "clinically relevant" for psychologists who might be called in to consult with athletes. Eventually they settled on 10 measures, including what they called dominance or leadership, aggression, achievement or drive, endurance or determination, emotional stability or control, mental toughness, coachability, trust, self-abasement, and conscientiousness. The test they designed presented "items," or statements, with which a test taker was supposed to agree or disagree, or about which they might remain neutral. So a prospect might answer "true," "false," or "uncertain" to the statement, "Hustle is important but it can't compensate for lack of talent." Such a response spoke to the trait of "achievement," helping scouts measure an athlete's "drive."[39]

The San Jose psychologists were careful to make sure the test was as robust as possible. They tested it with large groups of athletes, measuring its consistency (would specific items always point to the same characteristic?) and reliability (would the same person score similarly if he were to retake the test?). They included some items that had obvious answers to weed out those who couldn't read well or were careless in taking the test, as well as other items to try to pinpoint when players were just giving "socially desirable" answers rather than accurate ones.

The tests nevertheless had limitations from the start. Giving tests directly to athletes opened them up to manipulation, but few better options existed. Coaches, for example, were too inconsistent in rating athletes, with staff rarely agreeing among themselves on their own players' attributes. And when players ranked each other rather than themselves, only three traits seemed reliably measurable: achievement, self-confidence, and conscientiousness. Moreover, the tests crucially lacked "absolutely conclusive validity evidence" in the early 1970s and were inappropriate for female athletes and for athletes in individual sports (e.g., tennis and golf).[40]

Nevertheless the tests and their authors quickly found an audience hoping to efficiently discover which athletes had the "right" psychological profile. Tutko joined with Jack Richards to publish *Psychology of Coaching* in 1971, along with a companion "manual"—really just the AMI with instructions and interpretive guidance—and master pages for easy reproduction of the test. The manual even included a brief "Athlete's Guide to Athletic Motivation" meant to be copied and distributed to teams, which also advertised the Institute for the Study of Athletic Motivation and its experience measuring the personalities of thousands of athletes.[41]

In 1974 *People* magazine ran a profile on Tutko that asked him to address the "crisis" of young athletes increasingly focusing only on "money" and "winning." Tutko confirmed that the "preoccupation with 'winning only' and being No. 1 is a sick preoccupation for scholastic sports." His work had revealed that the professional athlete was "basically selfish—meaning his first thought is usually for himself." Tutko did acknowledge that there were some real role models in sports, people like Roberto Clemente, Henry Aaron, and Bill Russell. When asked whether coaches might use his tests' data against athletes, Tutko responded, "The purpose of the test is to set up a communication between the coach and athlete, to have them get together and discuss the results. Besides, authoritarian and holier-than-thou coaches won't touch the test. They think psychologists are part of the Communist menace, that we're going to take all the fun out of coaching."[42] There was still a lesson in all this, Tutko claimed, in that the best thing to focus on wasn't winning, but peak performance. That should be the goal of both coaching and psychological consulting.

The work of the Institute for the Study of Athletic Motivation was part of the expanding role that psychological expertise came to play in athletics in the 1960s and 1970s. In professional football,

Paul Brown brought psychological testing to bear on the selection of players for the Cincinnati Bengals after having used it earlier at the amateur level. The Bengals, along with other clubs including the Cowboys, Rams, and 49ers, relied on both the AMI and the more general Wonderlic Personnel Test in making decisions about player selection.[43]

By the 1980s a new guru for baseball psychology had emerged in the person of Harvey Dorfman, a consultant initially with Oakland and then with the Florida Marlins and the Scott Boras Corporation. Despite lacking an extensive academic background in psychology, Dorfman was part of the inaugural meeting of the Association for the Advancement of Applied Sport Psychology in 1986, at which he was joined on the baseball panel by Ken Ravizza, Ron Smith, and Karl Kuehl. The association's founding marked a shift in the role of sport psychologists, now formalized as a group that might be able to help clubs, and in the years since nearly every club has relied upon psychologists. In most cases, these consultants are hired to focus on player anxiety and tension. Though amateur sport psychologists continued to exist, the field had largely been professionalized by the 1980s, with journals, academic appointments, and research programs.[44]

Dorfman and other consultants during this period were more focused on peak performance of existing personnel than on the selection of new personnel through psychological profiling. Nevertheless, the promise of psychological tests as performance prediction tools lived on. By the late 1980s, the Sport Competition Anxiety Test, Sport Anxiety Scale, and Psychological Skills Inventory for Sport, among others, had replaced the Athletic Motivation Inventory test as leading predictive tools. In the 1990s, the Athletic Coping Skills Inventory was designed by researchers at the University of Washington in close collaboration with the Houston Astros' minor league teams. This test was meant to correlate with the Overall Average Evaluation, a single-score test incorporating eight-point scales of physical skills the organization used. This inventory was advertised as having greater validity

than previous tests, with more distinct factors and clearer connections between those factors and the test items. Its creators
found that in the minor leagues, physical and psychological components contributed about equally to position players' success,
but that psychological factors were far more important predictors
for pitchers' success. Scores on the test promised to reveal those
players whose psychology best predicted success in the game—
something better than just recognizing a kid's "makeup." Indeed,
as McLaughlin protégé Dave Ritterpusch explained in the 1980s,
sometimes baseball people may "be on to something, by intuition,
that's fundamentally important, like honesty or aggressiveness,"
when they talk about the "good face." "But how do you make very
fine distinctions?" he asked. "What kind of number-system would
you use to grade faces?"[45] That's where psychologists claimed to
offer something useful: a tool based not on intuition, but on science, one that produced a precise number, easily deployable in
calculations of prospects' value.

———

Clubs were constantly on the hunt for new ways of revealing and
analyzing the psychological and physiological attributes of prospects. The Phillies' 1978 manual noted that finding players with
"intense desire and determination" or "aptitude" was useful because these players, even if they had only "fringe" physical abilities, might still become "stars"; as examples, they cited Joe Morgan, Bob Bailor, Pete Rose, and Fred Patek. More to the point, the
manual noted that such players might elicit "little interest as free
agents," making an investment in them even more cost-effective.[46]

Such psychological findings were also useful in that "desire"
was a "great intangible," a basic quality necessary for the major
leagues. Being able to identify players without desire meant that
a team wouldn't waste time or money following them. This kind
of investigation required more than simple observation. Scouts
were instructed to befriend the prospect and to get to know his

family and teammates. They were also told to not be overly reliant on coaches, since they might be duplicitous, with other priorities in mind.[47]

Of course, that was not to suggest that scouts should try to do it all themselves. "Some scouts," the manual warned, "insist on playing psychologist and downgrade a player because they don't like a boy's makeup or appearance." The solution was to have scouts indicate their thoughts about a player's psychological makeup on reports, but not to let a distaste for a player's habits and dedication "influence" their judgment of his physical ability. By the early 1980s, it had become clear that the Phillies' front office didn't put much faith in physiological or psychological testing either. Jack Pastore, then the assistant scouting director, parodied the whole business: "I can show you an old psychological profile on Larry Bowa, where his 'guilt proneness' is rated above 90 and his 'emotional control' is below ten. That's interesting information, but how would you use it today if you were thinking about drafting or signing a kid like that?" Instead of such measures, what the Phillies told their scouts to determine was "how bad" a prospect wanted to play professional baseball. Scouting was just too complex to rely on only brute physical or psychological tests; there were "too many intangibles."[48] Instead, Philadelphia deferred to scouts' expertise—to their ability to know whether a player had desire and motivation.

The Phillies' resistance to psychological testing was largely a moot point—by this time, all top prospects in the nation were given a personality and motivation test, with scores distributed to each club. Players could decline, though that might not be interpreted charitably by potential employers. By the mid-1980s, in fact, there was wide agreement that clubs wanted the data even if it wasn't clear how useful they might be. McLaughlin, for one, believed he was able to correlate psychological test scores with a prospect's later achievement. If successful pitchers were both aggressive and emotionally controlled, those were traits that could be sought out. For Mike Arbuckle, a longtime scout and executive with Philadelphia, Atlanta, and Kansas City, "psychological testing is *a* tool. I don't think you ignore it, but I think it's just one more piece of information, as a Scouting Director at draft time, that you *may* want to factor in." Other scouts like Hep Cronin did not find such tests useful in evaluating players (it is always difficult to overlook an amazingly skilled player simply because a test suggests he might not be coachable), but many clubs still wanted the data because they could reveal useful information about a prospect's mindset, something that could perhaps be corrected or addressed by minor league staff.[49]

One common critique of psychological tests in baseball was that scouts were the ones who administered them, and those wishing to interpret their results had to watch out for "coaching." Scouts were sometimes tempted to help prospects they liked answer the items "correctly," ensuring that a player's desirability was not hurt by the test. Ritterpusch thought that unlike football and basketball scouts, who were college oriented, baseball scouts were far more skeptical of academic ideas like testing and scientific measurement. "It's like any other application of science," he said. "Your instruments are only as good as the people who use them. And you have to watch out for the fudge factor."[50]

Though tests' misuse was always a risk, the fact that scouts were their administrators also created opportunities. Cronin noted that the tests were "basically small tools," but "often more of an excuse for me to get into the kid's house and meet the family, to have some coffee and sit down and spend an hour-and-a-half or so with them. In that time, you also get a feeling for what the chances are that he will sign if you draft him." Lennie Merullo similarly appreciated that physiological and psychological tests got him "in the house," allowing him to talk to the family and learn more about the prospect than was visible on the field.[51] The context and setting of the test itself potentially provided scouts information, even if the results did not.

Scouts themselves were indeed often resistant to the idea that the tests might tell them something new. Former Cincinnati scouting director Julian Mock was such a skeptic. Noting that while many organizations gave psychological tests to players to determine their "mental makeup" before signing them, he was adamant that his teams didn't: "I feel we trained our scouts to find out answers to that stuff by using their own expertise. I don't like store-bought tests. They're just another tool for measuring scouts." Mock treated the tests like others did radar guns—they might produce detailed information, but it was of no better worth than what good scouts could find out using their "own expertise." To Lou Gorman, it took human intuition to figure out a player's psyche. "Things you can't see, how do you find those out?" he asked. "How do you decide? You can watch him in games, see what kind of drive he has in the ballgame, what kind of desire he has." But, he declared, the "best judge of a player is his peers, the guys who play with him. You don't fool them." A scout was a people person, someone who figured out which parent looked at his or her child through rose-colored glasses, or which coach revealed the truth about a player's desire.[52] Scouts' own expertise was in human relations, these critics seemed to suggest, and they hardly needed an outside expert to tell them whether a player had desire or fear.

Certainly the reformers pushing these new tools thought the scouts were simply resistant to scientific, objective approaches. When McLaughlin designed his "Whole Ball Player" chart in the name of science in the 1960s, it was in part because he believed then-current practices in scouting were woefully inadequate for the task of measuring psychological instincts. University of Delaware researcher Hannah thought the situation hadn't changed much in the subsequent decades. As he explained:

> The importance of our research wasn't so much the conclusions as the *numbers*. And if we'd been allowed to continue, it would have been possible, in several categories of physical ability, to establish cut-off levels for amateur players or at least to flag the poorer performers for close scrutiny before they were drafted or signed. . . . But the scouts don't see it that way. It's too much work. It's too technical for them.[53]

This was an old trope: critics had been complaining for decades that baseball failed to adequately take advantage of scientific insights.

Such conclusions are a bit unfair. Had physiologists and psychologists designed tests or measures that efficiently and reliably predicted success in professional baseball, it's hard to imagine they wouldn't have been adopted by clubs immediately. There was too much money at stake. Rather, as John Hoberman concluded back in 1992, "the much less dramatic truth is that 'sports psychology' is an eclectic group of theories and therapies in search of scientific respectability, as its more sophisticated academic theorists openly acknowledge."[54] Of course, such blanket dismissals ignored the ways in which psychological and physiological insights, from those focused on stretching and physical training to techniques of motivation and tools of measurement, had quietly pervaded the sport over the years. The truth was somewhere in the middle: scientific tools were neither useless nor essential for most ball clubs.

In fact, scouts' skepticism toward too-good-to-be-true mechanical shortcuts was shared by many scientists, who warned (and still warn) the public repeatedly that psychometric tests and tools shouldn't be used out of context. Coleman Griffith acknowledged in 1925 that "coaches and athletes have a right to be wary" because "there have been and are many men, short in psychological training and long in the use of the English language, who are doing psychology damage by advertising that they are ready to answer any and every question that comes up in any and every field. No sane psychologist is deceived by these self-styled apostles of a new day." Many of the efforts of academic psychologists early in the century were explicitly meant to replace the claims of "quacks" (as well as lay understandings) with more reliable, scientific procedures and insights.[55] Many of those within baseball who claimed the authority of psychology over the years had little formal training in the field, leading to the professionalization— and consequent policing of the boundaries—of sports psychology by the 1970s.

Moreover, even academic psychologists warned that as far as their tests had come by the end of the twentieth century, they were still not meant to be used by scouts and clubs to select among prospects. Even at their best, such tests explained less than half of the total variation in performance between players. Authors of the Athletic Coping Skills Inventory also cautioned in 1995 that self-reporting measures should "not be used for selection purposes, because they are too susceptible to response distortion."[56] Players—no less than scouts—could figure out what qualities teams desired and simply provide responses in line with those qualities. Nevertheless, their use continued, in large part because they remained a relatively cheap way to reduce psychological character to a number.

The role of psychology in baseball scouting was just one part of a broad expansion of psychology into domains previously dismissed as inescapably subjective. Like McLaughlin and Hannah, some psychologists came to see subjectivity as only an obstacle to be overcome. Much of postwar psychology was devoted to creating objective, especially numerical, accounts of subjective attitudes, desires, and thought processes. The point was never to eliminate judgments but rather to find new ways of expressing them numerically.[57]

———

Science has promised a more rational approach to playing baseball for so long now that the schemes themselves are sometimes repetitive. When modern teams position players defensively on the basis of route efficiency data or spray charts, for example, they are updating one of Evers and Fullerton's proposals from a century earlier. Evers and Fullerton in *Touching Second* had plotted the initial position of fielders and then calculated the range for an outfielder given that an average fielder could cover 50 yards in 6 seconds and that the ball was in the air for about 3 seconds; for an infielder, the ball would get to him in about 1.5 seconds.

They even took into account the fact that most outfielders would have a "fast" and a "slow" side in moving toward the ball, so their trajectories wouldn't be symmetric. The point was to show the "grooves" in which batters might want to hit, as well as the optimal defensive positioning.[58] Fullerton's subsequent article on Babe Ruth's psycho-motor evaluation and Hannah's research program nearly a half-century later both suggested that if only bat speed and acceleration, response time and perception, could be measured accurately, then those with superior but possibly latent abilities could be isolated, cultivated, and exploited. In 2006, the men's magazine *GQ* persuaded Albert Pujols, then a reigning batting star, to undergo the *same* tests that Ruth had done in 1921. Like the original, this was clearly a publicity stunt, but it was also meant to show that baseball's ways still needed correcting. The article approvingly quoted a trainer who noted that the Ruth tests remained "more appropriate and more intensive" than any other current testing regime. The author, Nate Penn, concluded that baseball "needs to adopt an unsentimental, scientific approach to player evaluation and training." (Though not every test could be replicated, Pujols bested Ruth in tests of bat speed and motor control, and was far above average in every category tested.) Penn also echoed the University of Delaware research program in promoting the value of the measurement of depth perception and highly developed binocularity and suggested a successor to the AMI as a way to gauge a prospect's psychological profile. Even the cast of characters was familiar: Ritterpusch, the protégé of McLaughlin in Baltimore, was quoted still pushing the idea that psychometrics was the key to predicting player success.[59]

Dreams that ever more powerful analyses are possible—if only the data can be collected—remain pervasive. Steve Mann, a creator (alongside Dick Cramer) of STATS and a pioneer in the real-time gathering of baseball data, also remains a believer that science is on the brink of unlocking players' full potential. At the 2013 SABR Analytics Conference, Mann applauded the fact that front offices were becoming ever more scientifically savvy and could now draw upon new research done on bat speed and swing power generation. Similarly, in 2016, the *Washington Post* ran a story noting that two neuroscientists had recently published work that might provide the key to understanding batters' thoughts as pitches approached the plate. This study found that certain hitters may simply excel at inhibiting their swing, and finding out who they are might enable clubs to predict success. (The scientists have even developed an app that markets itself as improving players' pitch recognition.) Two years later, journalist Zach Schonbrun updated that research in a *New York Times* article proclaiming "Baseball Really Is Brain Science." Companies are peddling "wearable technology" that promises to warn teams before injuries occur, as well as "neuroscouting" tools that purport to offer new ways of sorting the mentally strong from the weak. Many scientific enthusiasts continue to see advances in research, alongside Statcast and other data collection efforts, as avenues to make player acquisition and development more rational.[60] While such technologies are new—as is the amount of data collected—the basic premise that clubs might use science to provide objective measures of physical and psychological states is not.

The hopes of Ritterpusch, Hannah, and McLaughlin nevertheless remain on the periphery of the game as it is played, even in the twenty-first century. Penn, the author of *GQ*'s article on Pujols, offered the explanation that baseball guys reject such ideas because they're "faithful to the stale myth of divining talent." This conclusion echoed Hannah's complaint that baseball people were "forever steeped in the traditional." In this account, scouts' devotion to subjectivity was the problem. Scouts, Hannah said, were "*proud* of being subjective. So if you wanted to apply technical testing to scouting, you'd have to hire someone more qualified than scouts to do the testing."[61] At the same time, he dismissed scouts' claims to expertise, concluding that subjective knowledge can never be expert knowledge.

Scouts, however, have long acted as if subjective judgments *can*—given expertise and the right tools—be turned into reliable knowledge. "Subjective would be if you wanted to sign a boy because you were swayed by the enthusiasm of his coach," Brandy Davis explained. "Or if you put the dollar sign on the muscle and then raised it just because you thought he was such a nice kid—we call that 'falling in love' with a boy. . . . But 'good face' is *ob*jective: it means he impresses you as an athlete—not [as] a pretty boy." Objectivity here was about accurately assessing body types and qualities. In these assertions scouts echoed themes from the long history of phrenology, craniometry, physical anthropology, and other biometric sciences, as well as the somatotypes of psychologist William Herbert Sheldon from the 1940s. The language of the athletic "type" or "temperament" is old indeed, going back at least to the nineteenth century, when scientists first started to investigate whether those who excelled at sports were not simply people who happened to possess particular skills but different sorts of people altogether.[62] Perhaps some people were natural athletes, and it might behoove those in charge of a team or sport to find methods of filtering out everyone else. Scouts were not rejecting objective measures when they expressed skepticism about radar guns, stopwatches, and psychological tests. Instead, they were suggesting that objective evaluation of prospects was about more than clocking a fastball or scoring a psychological test. It required accurate assessments of prospects' physical and mental types and judgments about those types' future development. Moreover, as the next chapter will show, scorers increasingly expressed those assessments and judgments with numbers.

Radar guns, stopwatches, psychological tests, ortho-raters—all have been pushed as shortcuts to revealing whether prospects have otherwise invisible characteristics that will ensure or prevent their success. They all have the virtue of expressing their measurements numerically. But none have replaced the expertise of scouts in the evaluation of bodies. Scouting hasn't persevered because scouts somehow have enough clout to keep teams from trying to find cheap shortcuts. Scouts don't have that much power. Rather, scouts have survived because teams have never found any device that can replace their expertise.

7

A Machine for Objectivity

Whatever they thought about a prospect, scouts usually had to express their judgments numerically. Phil Rossi, working for Boston in preparation for the 1987 draft, wrote that Craig Biggio was "strong," with a "sturdy ath[letic] build"—in short, a "quality player" with "potential." Rossi also precisely specified this overall potential as 66.0 (on a 20–80 scale) based on the three games in which he had seen him play. Brad Kohler's report rated Biggio's potential as 65.5, noted he was a contact hitter with only average power, and recommended him as a "very agg[ressive], take charge type player." Don Labossiere, scouting Biggio a year earlier for San Diego, was not as convinced. He measured his potential as 46, noting that Biggio did not yet have a major league bat or power but remarking that he had good "makeup" and the potential to improve.[1]

These reports relied on impressively precise numbers, with values calculated down to the dollar, ratings calculated to the tenth of a point, and grades specified for power, running speed, fielding ability, aggressiveness, and at least six other factors. Scouts managed to translate the messiness of evaluation into single numbers: "dollar value" and "overall future potential." These numbers were useful because teams needed to rank all prospects in preparation for the draft. Knowing there were two excellent pitching prospects was not enough—one of them had to be ranked above the other because one of them had to be selected first.

Modern scouting was fundamentally changed by the institution of the major league draft in 1965. After the draft was initiated, general managers, scouting directors, and scouts themselves had to learn how to make scouting reports useful in the evaluation and ranking of players. Clubs joined together to form the Central Scouting Bureau and later the Major League Scouting Bureau to consolidate scouting efforts. As the nature of scouting transitioned from signing players cheaply to accurately rating players, numerical modes of evaluation became increasingly prized for their ability to be standardized and compared. Signing players went from largely a one-on-one negotiation to a bureaucratic, multilayered process. Detailed instructions for how to fill out elaborate reports were published. Within two decades of the draft, nearly every club used an identical scouting form.

Scouts did not try to turn subjective judgments into objective evaluations in the way that psychological tests reduce complex phenomena to numbers. Rather, scouting knowledge gained reliability and objectivity through bureaucratic management. Scouts' expert judgments never went away but were increasingly incorporated into a structure through which reliable, numerical judgments of players could be made. Numbers never replaced or stood in for scouts but rather became part of their practice. By the 1980s, scouts had put a single number on Biggio's potential because that was the primary way they communicated his value to the club and to each other.

———

Joe Branzell was just starting as a scout in 1961 when a new Washington Senators club was created to replace one that had moved to Minnesota. This was one of the first expansion teams since the National League–American League merger in the early twentieth century, and the organization had to be built from the ground up. Branzell joined a staff of 24 full-time scouts, a general manager, an assistant general manager, and a "chief" scout and was put in charge of scouting the Maryland–Washington, DC, area.

Washington's manual for scouts in 1961 explained Branzell's duties. After making the obvious point that his primary job was to find and sign the best players possible, much of the manual focused on logistics. Pages and pages were needed to explain eligibility guidelines, the difference between contingent bonuses and deferred bonuses, age restrictions, and tryout rules. Anyone scouts were hoping to sign should have been carefully cultivated over time—letters every few months, phone calls as scouts passed through prospects' hometowns, and so on—to generate "goodwill" that would allow scouts to sign players without complication. Branzell would have to contact the Senators' central office to find out minor league roster needs before signing a player—as well as the maximum amount he could offer the player.[2]

The Senators also provided a series of hints for signing players, because "all of the scouting in the world is not worth anything unless you are successful in signing the player": don't leave written offers after making them (either get the player to sign them or take them back, else they be used as leverage with another club), for example, or don't discuss terms until the player is ready to sign. The Senators even suggested the shape scouts' questions might take: "What would I have to do to sign you right now?" "Would you consider an offer of 'X' number of dollars?" Finally, if a player hesitated, scouts were encouraged to offer up some suggestions, selling the prospect of the Senators as a young organization with the opportunity of rapid advancement (an argument better made by bad teams and new teams than ones stocked with famous players). Joining a new organization meant a prospect would soon be known by all the people who "counted," scouts were told to say. Washington, DC, could be "sold as a pleasant place to live," to boot.[3]

This advice was ultimately about money. Selling the city and club to prospects was a way of encouraging players to sign for less money than they might have been able to command from another organization. Cultivating goodwill was central to signing players without much competition and therefore with lower bonuses. Some players, especially those recently released by another organization, were never offered bonuses, on the basis that a second chance to play should be incentive enough. Scouts in the early 1960s were practitioners of the time-honored trade of signing players as cheaply as possible.

The manual also gave instructions for basic accounting practices. Scouts were responsible for logging weekly reports indicating what games they saw each day and the progress of players under consideration. Scouts were also supposed to indicate when players they had seen were signed to other clubs, noting what bonus had been offered and what the scout himself had thought about the player. Was the other club making a mistake? Had the scout overlooked a prospect? Such recordkeeping was a way for the club to track its investments in both players *and* scouts.

Much of the activity of scouting—if not its core purpose—was in public relations. A separate section in the Senators' manual on that topic indicated that it was the responsibility of the scout to keep contact with media outlets, coaches, and other interested parties. Scouts needed to sell the club so that local prospects would have a positive image of the Senators as well as promote its signings in local media outlets. Senators' fans should know that the signed players had not only physical skills but also character and desire. The Senators wanted the scout and the club, not the player, to provide the narrative, which was that prospects were joining a "young, progressive, imaginative, and enthusiastic organization."

Scouts themselves were responsible for appearing "neat" at all times—a tie and suit or sports jacket was required when visiting media offices or when meeting with any baseball organizations. Scouts were told to remain clean-shaven and avoid off-color jokes. The scout was meant to model the mores of the organization itself. Whether this happened in practice was a different issue—Branzell, for one, was so economical that he washed his clothes in hotel sinks and carried around cooking equipment as he traveled. A colleague once pointed out that he needed a tetanus shot to get into Branzell's car.[4] Economy was certainly a cardinal virtue for Branzell and the Senators—if a scout had done his job, then he would be able to sign prospects cheaply, with little risk to the club.

The business of scouting for the Senators relied upon "player information cards," or scouting reports. One half of the card required a narrative paragraph, not that dissimilar from the informal letters and telegraphs that had been passed between scouts and the front office for decades. The paragraph described the physical and mental condition of the prospect, from specific details of his throwing motion ("throws straight overhand") to generalizations about his motivation ("competitive spirit appears good"). Issues like signability ("may have trouble in college") or areas for improvement ("may have to bend his back more") were also noted. The point was to concisely provide "general impressions of the player as well as specific elaborations about individual skills." That is, the paragraph was meant to provide a thumbnail impression of the player's overall abilities and prospects for improvement. In 1961, at least, the manual warned that scouts were often "lax" in this type of reporting and encouraged them to improve.[5]

The Senators' manual also gave guidelines for what might or might not be worth noting. In particular, the club prized young, big, fast, and strong players over other types. There were "exceptions," of course, that required individual judgment, and—the manual pointedly reminded scouts—judgment was "the thing for which you are being paid."[6]

The other half of the card (and the whole report was, physically, just two sides of an index card) required biographical information and numerical grades. Scouts were instructed to evaluate specific skills for position players (arm and accuracy, fielding and range, running and base running, hitting and power) and pitchers (speed, curve, change, control) as well as aptitude and aggressiveness for all players. Each item was to be assigned a number between 1 (very poor) and 7 (excellent).[7] The grade was meant to be given on the basis of major league ability (with 4 being average). No plus or minus signs were allowed, though the club would tolerate decimals (4.5) if the scout insisted. The final judgment to be provided on the card was a "yes" or "no" to the question of whether the player was ultimately a "definite prospect" for playing in the major leagues.

By 1961, as Branzell began his career with the Senators, most of the modern apparatus of scouting was in place: descriptions of a body and its faults, numerical grades of skills, narrative judgments about a player's prospects. The report was still largely informal, however, in that it was effectively a private document shared between the scout and his supervisor, more a business record of a transaction than the basis of a larger dataset. That is, the report was not meant to be widely compared with others. Though an organization's signings were obviously interconnected—there was only one budget—the player information card system treated each transaction individually. The scout reported on the players he saw, whether they were signed, and what they were ultimately worth. Branzell would soon, however, find himself learning an entirely new system as the world of scouting shifted decisively around him.

This shift was the result of the "rule four" draft, instituted for the first time on June 8, 1965. Over the 1950s, baseball clubs became increasingly tired of paying top prospects ever more money in signing bonuses.[8] Nearly all top players knew that they could entertain multiple offers, playing organizations against each other. The leagues' initial approach to this problem was to implement various versions of a "bonus rule," which effectively forced teams that paid out high bonuses to prematurely use one of their major league roster spots on the signee or risk losing the player through the so-called minor league or "rule five" draft. Such mechanisms proved largely ineffective at keeping deep-pocketed clubs from stockpiling all the top talent (and were rife with charges of under-the-table payments and a lack of enforceability). Though the minor league draft was maintained, the most punitive elements of the bonus rule had effectively been eliminated by 1957.

Few owners thought bidding was the best way of hiring players, though some were happy with the system, particularly wealthy organizations that benefited from the ability to sign prospects largely at will. The status quo therefore also suited the expansion clubs of the 1960s, with their penchant for lavish spending on marquee players. Like Branzell's Senators, the expansion clubs of the era were located in major markets—Los Angeles, Houston, and New York—and owners were willing to spend money to build a fan base. These clubs (except for Houston) and the wealthiest clubs would become the core dissenters to the draft plan.[9]

In the 1964 winter meetings, 13 of the 20 clubs approved the creation of a major league draft as a cost-control measure. The owners of these 13 clubs expressed fear that without a draft the competitive imbalance that had prevailed in the 1940s and 1950s—the "golden era" of baseball for fans of the three clubs that won 15 of the 20 world championships before the draft was instituted—would continue. (All three predictably opposed the draft.) Cleveland's general manager, Gabe Paul, made the case for competitive balance when he explained his support of the draft: "It'll help equalize our teams. We have to do something to give the weaker teams greater incentive" and address "player procurement and development."[10] Baseball was the last of the American professional sports to institute a draft, following football, basketball, and hockey.

The basic mechanism of the baseball draft has remained the same since 1965: Teams select prospects in the reverse order of the previous year's standings. Once drafted, players are restricted to that team for a certain amount of time (the length of which has fluctuated over the years); initially, clubs' draft picks were tied to roster spots at specific minor league levels (AAA, AA, etc.). In addition to the main June draft, there was originally an American Legion draft in August, but this was quickly discontinued. Other modifications have been made over the years (concerning the eligibility of college students, Canadians, and Puerto Ricans; the addition of "supplemental rounds"; and so on), but the basic format remains the same. Each team initially has one pick in each round and then exclusive rights to try to sign that player until he is eligible to be drafted again.

These exclusive rights were meant to keep bonuses down—players who didn't sign had to forgo their salary while they waited for the next draft. Underlying the draft was the presumption that the best players around the country were known and should be parceled out in an equitable manner rather than allowing the richest teams to engage in a bidding war. That didn't mean there was universal agreement about the best players, of course—predicting success has never been easy—and in the first draft Johnny Bench was ignored the entire first round, and Nolan Ryan for the first 11 rounds (i.e., for 294 players). Both would have Hall of Fame careers. Even if there were players who were drafted higher than expected and thus were able to demand more money than they might have gotten in the old system, bonuses on the whole were kept controlled. Despite the dawning of "free agency" in the 1970s and the rapid increase in salaries for veteran players, only in the late 1980s did signing bonuses regularly top $205,000, passing the nominal price at which Rick Reichardt was signed by the Angels in the last predraft year of 1964.[11]

For scouts who worked around the time of the draft's institution or shortly thereafter, the draft was a killjoy. P. J. Dragseth's 2010 collection of interviews with veteran scouts, *Eye for Talent*, can sometimes read as if it were a prolonged therapy session for scouts to vent about the draft. Ellis Clary told Dragseth, "Scouting is no fun anymore since that draft thing started." Jerry Gardner echoed the sentiment that the draft "really took the fun out of" scouting, explaining that "it's more reporting now." Gene Thompson offered, "In my opinion the draft was a big mistake." Phil Rizzo waxed nostalgic about the "competition" and "bidding wars" before the draft. He found that the draft confused everyone: It's "just a farce. It doesn't work the way we were told it was supposed to. I don't think scouts today get it, and I don't think half of the people in baseball get it."[12]

At first, neither scouts nor front offices knew how to "do" the draft. Joe Brown, Pittsburgh's general manager in 1965, claimed he went into the first draft "blind," with just a list of 450 names and trunks full of player notes. Hugh Alexander, whose scouting career spanned the draft, noted when it took place: "I didn't know how to handle the draft and neither did the people who ran the big league clubs."[13] These were typical responses—the draft fundamentally changed the way clubs acquired talent. And some didn't like the change at all, with the most outspoken likely being Philadelphia's National League club. As club officials told their scouts, the Phillies were "fundamentally opposed to the Free Agent Draft and its inherent socialistic qualities."[14]

Whether or not it was "socialistic," the draft certainly created a sense of artificial agreement among a club's scouts—at least publicly. Before, each scout had worked in his own area and was largely able to sign players independently. Now efforts had to be cooperative; there were not separate drafts for each region, but one for the country. Area scouts' reports had to be combined to create a single draft list, with players ranked one-by-one.

Consequently, the draft revived the idea of communal scouting efforts. Ever on the lookout for cost-saving measures, Branch Rickey had recommended "pooled scouting" in his 1965 book *The American Diamond*. (Jim Gallagher was later also credited with publicizing the basic idea of pooled scouting back in 1962, during his time as the Chicago Cubs' general manager, though the concept had few supporters then.) Rickey's idea was for the pooled scouts to be highly skilled and therefore able to sift through prospects and create a master list of about 500 or so players who would be available for drafting by individual teams. Rickey had also suggested that a central scouting bureau be created as part of his planned Continental League, but when its creation was made moot by the expansion of the major leagues, the idea was set aside.[15]

After the draft, the calculus for clubs changed. Since each club now had a chance to draft every player, and rich ones couldn't corner the market, spending lots of money on loyal scouts who might "discover" a hidden gem no longer gave the same return on investment it once did. Chances were good the player would already be known to another team that happened to have an earlier draft pick. In 1965, Pittsburgh, Milwaukee, Cleveland, Chicago, and Detroit made an informal agreement to share scouting resources, and soon the commissioner's office initiated plans for a centralized bureau.

Three years later, Milwaukee's Jim Fanning was put in charge of the creation of the Central Scouting Bureau. Building on the model of shared resources that Pittsburgh and others had created, Fanning saw the organization not as a replacement for clubs' scouting departments but rather as an additional source of information on the huge number of players who would be drafted after the first few rounds. As Fanning knew from his experience as a farm director, it was difficult for clubs to decide whom to select after the top 100 or so prospects were taken. For the relatively low annual cost of $7,500, the bureau would provide information to clubs on these later rounds so that general managers weren't just "picking names."[16] The bureau began scouting only in Florida and the Dakotas as a proof of concept, with plans to expand to the Rocky Mountain and New England regions and eventually nationwide.

The pooling of individual judgments required more than just the creation of a procedure for mailing reports to all 20 clubs. It also meant standardizing the judgments of bureau scouts. These scouts would have used all sorts of different forms in their previous jobs, with a range of instructions about how to grade players and report those evaluations. Now they needed to produce uniform information and reports that were calibrated and therefore comparable across every one of the nation's prospects. Fanning produced model reports for bureau training purposes in 1968 to show how he wanted scouting reports to be standardized. Each report had one hole punched in the upper right to indicate position (1B, OF, etc.) and one in the lower left to indicate prospect quality (definite, good, etc.). This way a club could quickly isolate all first basemen or all definite prospects in preparation for the draft. Each card then standardized ratings (with 1 being "excellent" and

5 being "poor") for batting, power, speed, running, fielding, arm, and "other." Each also provided stock phrases for descriptions of players' bodies and skills ("body control," "good wrists," "desire," etc.), which scouts only needed to circle. This forced scouts to standardize referents, enabling comparison of players regardless of the person scouting. There were places to indicate race (W, N, Latin) and special circumstances (glasses, single, married, military draft status). Each club would be able to learn how to read and interpret the report even if it did not know the scout himself. Bureau reports thus standardized judgments about prospects.

The Central Scouting Bureau didn't last long—Fanning left within six months to become the new general manager for Montreal's expansion team, and his successor, Vedie Himsl, carried on for another season but effectively dissolved the bureau in 1970. Part of the problem was that clubs were reluctant to give up their best scouts, ensuring the bureau's reports were of lower quality than clubs' own assessments. But it was also because there was not sufficient buy-in.[17]

The idea didn't die, however, and by 1974 there was a new movement for a centralized scouting system. Joe Brown, who was still general manager of the Pirates, again pushed for the revival of a coordinated scouting bureau in the 1970s. Joining him were Lou Gorman of Kansas City, Spec Richardson of Houston, Harry Dalton of Baltimore and California, and American League president Lee MacPhail. Their main argument was that labor costs were increasing and as some form of free agency seemed ever more likely for players, it was clear that other expenses would need to be reduced. Every American League club except Chicago and every National League club except the expansion clubs in San Diego and New York and the big-market Los Angeles, San Francisco, St. Louis, and Philadelphia clubs paid nearly $120,000 to restart the pooled scouting system, now under the name of the Major League Scouting Bureau.[18] Jim Wilson, Milwaukee's general manager, was hired to head the new bureau alongside Don Pries, Baltimore's assistant general manager.

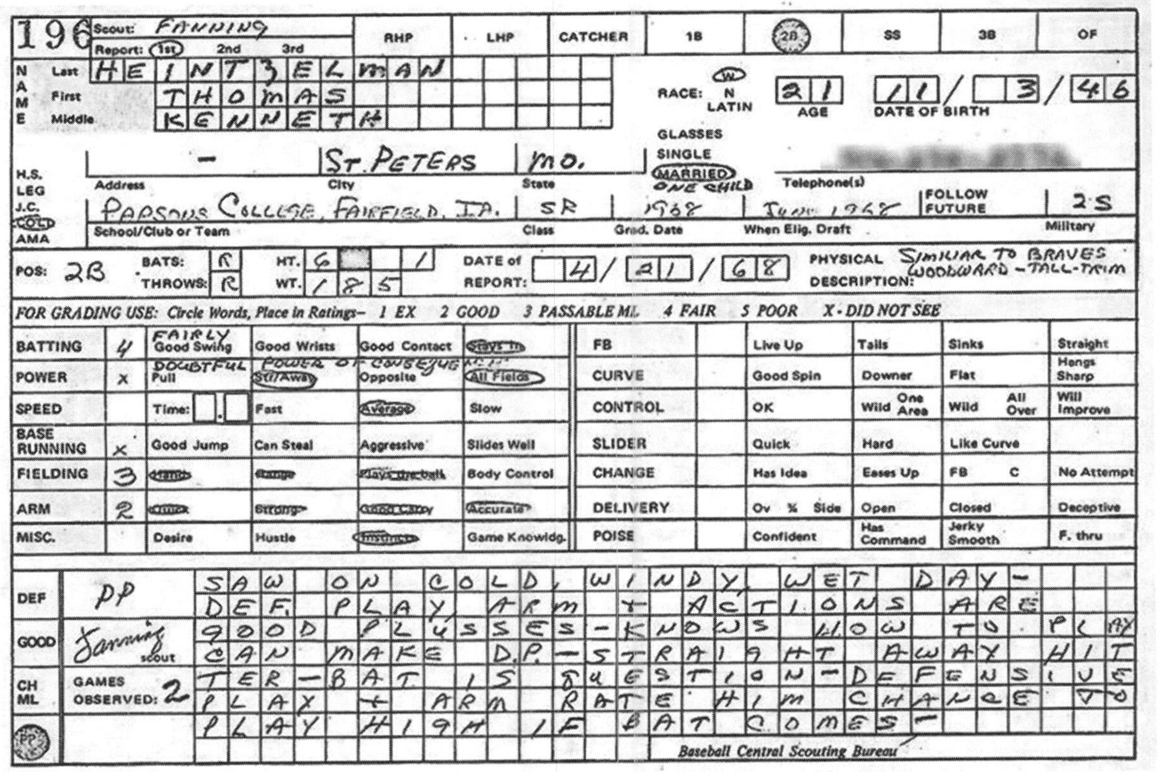

196 Scout: FANNING Report: (1st) 2nd 3rd

RHP | LHP | CATCHER | 1B | (2B) | SS | 3B | OF

NAME Last: HEINTZELMAN
First: THOMAS
Middle: KENNETH

RACE: (W) N LATIN
AGE: 21
DATE OF BIRTH: 11/ 3/46

GLASSES
SINGLE
(MARRIED) ONE CHILD

H.S. / LEG / J.C. / (COL) / AMA

Address: —
City: ST. PETERS
State: MO.
Telephone(s)

School/Club or Team: PARSONS COLLEGE, FAIRFIELD, IA.
Class: SR
Grad. Date: 1968
When Elig. Draft: JUNE 1968
FOLLOW FUTURE
Military: 2S

POS: 2B
BATS: R THROWS: R
HT. 6 1 WT. 185
DATE of REPORT: 4/21/68
PHYSICAL DESCRIPTION: SIMILAR TO BRAVES WOODWARD – TALL-TRIM

FOR GRADING USE: Circle Words, Place in Ratings– 1 EX 2 GOOD 3 PASSABLE ML 4 FAIR 5 POOR X - DID NOT SEE

BATTING	4	FAIRLY Good Swing	Good Wrists	Good Contact	Stays In	FB		Live Up	Tails	Sinks	Straight
POWER	X	DOUBTFUL Pull	POWER OF CONSEQUENCE Str/Away	Opposite	All Fields	CURVE		Good Spin	Downer	Flat	Hangs Sharp
SPEED		Time: .	Fast	Average	Slow	CONTROL		OK	Wild One Area	Wild All Over	Will Improve
BASE RUNNING	X	Good Jump	Can Steal	Aggressive	Slides Well	SLIDER		Quick	Hard	Like Curve	
FIELDING	3	Hands	Range	Plays the ball	Body Control	CHANGE		Has Idea	Eases Up	FB C	No Attempt
ARM	2	Quick	Strong	Good Carry	Accurate	DELIVERY		Ov ¾ Side	Open	Closed	Deceptive
MISC.		Desire	Hustle	Instincts	Game Knowldg.	POISE		Confident	Has Command	Jerky Smooth	F. thru

DEF: PP
GOOD: Fanning, scout
CH ML: GAMES OBSERVED: 2

SAW ON COLD, WINDY, WET DAY –
DEF. PLAY, ARM & ACTIONS ARE
GOOD PLUSSES – KNOWS HOW TO PLAY
CAN MAKE D.P. – STRAIGHT AWAY HIT
TER – BAT IS QUESTION – DEFENSIVE
PLAY & ARM RATE HIM CHANCE TO
PLAY HIGH IF BAT COMES –

Baseball Central Scouting Bureau

FIGURE 7.1: Jim Fanning's 1968 report on Tom Heintzelman was one of the earliest Central Scouting Bureau reports, showing how holes were punched to quickly sort by position and prospect level. Scouting Report Collection, BHOF

Wilson and Pries envisioned the bureau operating like the scouting department of a traditional club—that is, with "outstanding coverage" of the country provided by "high caliber" scouts. Gordon Lakey was appointed as an administrator for the bureau, which planned to hire 40 full-time and 30 part-time scouts. Wilson wanted the bureau to be the largest and richest scouting operation in the nation. As the bureau eventually explained in its manual, the goal was not only to provide more stable employment for scouts, but also to make scouting more effective, productive, and efficient. Offering good salaries and benefits, the bureau was able to sign 56 scouts within three weeks of beginning operations, and other clubs were forced to raise their own benefits to match or risk losing their top talent.[19] Scouts may have continued to think of themselves as loners, but they were becoming ever more like employees of typical corporations, negotiating benefit packages and job responsibilities.

By the end of 1976, only five clubs remained nonmembers of the bureau—the Toronto expansion club, as well as the holdouts in Los Angeles, Philadelphia, St. Louis, and San Diego. As a result, the yearly fee for a club's participation had been reduced to just a shade over $80,000. But when the initial three-year term was up in 1977, clubs began dropping their membership, and the percentage of member teams fell dangerously close to 50 percent by 1980. Some members suspected clubs were skipping out on the fees only to cheat by accessing reports under the table—San Diego in particular appeared to be receiving the same scouting information as paying members, critics insinuated.[20]

Bureau members consequently wanted to make the fees mandatory for all clubs, and in the 1983 winter meetings, which focused on the replacement of Commissioner Bowie Kuhn, 15 clubs (a majority) were able to force through mandatory membership in the bureau. This not only preserved the bureau but ensured fees would be kept low and leaks would not hurt subscribers.[21] Serious questions about the bureau's usefulness remained, however, especially after a season in which all four playoff teams were nonmembers. Nevertheless, with the arrival of Peter Ueberroth as commissioner in 1984, the Major League Scouting Bureau was put under the aegis of the commissioner's office and made permanent.

Clubs had different ways of using the bureau. Milwaukee, under Bud Selig in the late 1970s, had relied heavily on the bureau but by the early 1980s used it only as "an insurance system, a safety net," for ensuring no top prospect was overlooked. Other clubs, like St. Louis, came to use it as an initial screening tool—their own scouts were supposed to see anyone who received a decent evaluation from the bureau, a requirement that some scouts resented as just creating "more work" for them. The bureau certainly did save clubs money. The *Sporting News* counted 168 scouts fired in the year after the bureau's creation, though 67 of them were quickly hired back by the bureau itself. Within a decade, teams had divided into camps: some depended only on the bureau or on the bureau plus a skeleton scouting staff, others combined a robust staff with the bureau's reports as additional data points, and a final group simply ignored bureau reports and used their own scouts. By the early 1980s, these choices were largely about how much money clubs wanted to spend on player acquisition, ranging from well under $1 million for bureau-only clubs to nearly $2 million for clubs with large staffs.[22]

———

Even if the bureau was a mixed bag in the minds of contemporary observers, the organization definitely transformed the practice of scouting. By 1975, Branzell's Senators had become the Texas Rangers, and the Major League Scouting Bureau's manual had become the club's own scouting manual, which the Rangers simply customized by annotating the original and making copies. The bureau's objectives and loyalty oath were still included, though the Rangers had typed a supplementary note in the manual indicating that the flow of information was "one way," from the bureau to the Rangers.[23] The club's scouts were loyal only to the Rangers, not to the bureau, but bureau scouts could be called upon by anyone. By relying upon the bureau, the club had reduced its scouting staff to just 13 from over two dozen but also listed the 60-odd bureau scouts as if they too worked for the Rangers. Scouting for the club was conceived as a joint effort.

In 1975 the Rangers told scouts like Branzell that their role was now to accurately fill out scouting reports. Even opponents of the draft and the bureau, like Philadelphia, told their scouts that "scouting reports are an absolute necessity if the job is to be done properly." They might be "very time-consuming," the club noted, but "their importance cannot be overemphasized particularly in view of the Free Agent Draft." These reports were "referred to constantly, and in many cases, are vital in the making of trades." Similarly, the New York Yankees made it clear that after the draft, scouting was "no longer strictly an individual effort." Now, scouts needed to see as many prospects as possible, evaluate them "properly," and "convey the information to a centralized point where it may be correlated with similar information from scouts from other areas."[24] After the draft, and particularly after the creation of the Major League Scouting Bureau, Branzell and other scouts' job shifted from signing players to participating in a large, standardized data-gathering effort.

Branzell and the Rangers also adapted the bureau's "Free Agent Report" for their own uses. This move had strategic value in that by effectively using the same report it was easier to integrate player analysis. Front office personnel didn't want to reconcile two different kinds of reports on a player—they wanted all information to be comparable. The Texas club did in fact print separate reports from the bureau, with a carefully annotated guide on how to use them, but the format was almost exactly the same as that of the bureau's own reports.

Wilson and Pries had made the assignment of grades ranging from 2 to 8 for each skill the core of the bureau's report. Rangers' scouts were consequently asked to assign to each player's performance in a category (range, fielding, base running, power, etc.) a whole number between 2 ("poor") and 8 ("outstanding")—with 5 being average—based on "major league standards, not amateur." That is, each player's skills were compared to current major leaguers and expressed numerically.

The use of numbers was ultimately about increasing precision, an aim all clubs desired to meet once they were responsible for comparing every available prospect in preparation for the draft. In the 1940s and 1950s, it was common to simply distinguish exceptional prospects from fair ones—after all, in both cases the important question was whether they were worth the money it would take to sign them. Pittsburgh's manual for scouts in 1949 exemplified this approach, offering the advice to label each player using the "code" of "exceptional," "good," "fair," or "poor."[25] That's all the front office really needed to know—in addition to the price each prospect was worth, of course. The draft required more precision. Simply knowing a player was a "fair" prospect didn't help clubs distinguish him from thousands of other marginal players.

Though numerical grades were not a product of the draft, the draft made them essential. Al Campanis, a scout trained by Branch Rickey and an influential figure in midcentury front offices, claimed that he was an early advocate of the move from words (average, average-plus, etc.) to numbers, in search of something more "refined." Campanis compared scouting to grading in school, describing the scout as a "schoolteacher," and the system one in which "70 is a passing grade, so that can represent the major-league average on arm or speed or whatever, and 60 and 80 can be the extremes." So whereas before a scout might have had a hard time distinguishing a slightly weak arm from an average one in his reports, now he could give it a 69, and "anybody reading your report would know you were talking about a kid who still could become a major-league second baseman."[26]

Campanis's desire for "refined" and precise grades in order to facilitate communication was widely shared. Philadelphia, for example, told its scouts in 1972 that the club was moving to a numerical scale to make its evaluations more "precise." The old system was "too vague and contributed to a breakdown in communications between the scouts in the field and the front office." The Phillies, pushed by scouting director Brandy Davis, adopted Campanis's 60–80 scale and encouraged scouts to use numbers other than 65 or 75—the point was precision, so they should "give one grade only" and "not generalize."[27]

Not everyone adopted Campanis's 60–80 scale, however. Howie Haak of Pittsburgh used a 0–60 scale, with 30 as a major league average. This scale functioned similarly to Campanis's, though it offered even more precision—60 gradations instead of 20. Other clubs used 10-, 7-, or even 5-point scales. Whatever scale was chosen, Campanis concluded that "a number system of *some* kind is the standard vocabulary today, because the nature of modern scouting is that you have to communicate your grades to more people."[28] He explained that numbers enabled scouts to make their judgments increasingly precise, and that made it easier to communicate numerical judgments reliably to others. He had faith in the power of numbers to convey information efficiently—people knew what it meant to give a 69 instead of a 70.

After the Major League Scouting Bureau began, nearly every club adopted its scale, with individual ratings between 2 and 8, and 5 as the major league average. (This particular range was perhaps modeled on the increasingly popular SAT, which by midcentury used a 200–800 scale for each section and set 500 as an average. In any case, no one seems to remember its specific origins.) It was still possible, of course, to figure out correspondences between reports, and some clubs indeed maintained their own systems for years. Nevertheless, the convergence wasn't surprising. For clubs using the bureau's reports, there was a strong incentive to make their own documents commensurate with them.

In addition to its promotion of the 2–8 scale, the bureau was central to the spread of a distinction between "present" and "future" grades. Scouts' grades were most relevant for what they thought the player would become, not who he was now. For each skill, scouts were required not just to report on what they saw but to use that observation to predict future ability. The bureau provided its first scouts the example of Assistant Director Gordon Lakey's May 1974 report on Kevin Drake. Some of Lakey's projections in this report were quite dramatic: Drake currently hit at the 2 ("poor") level but was projected to hit at a 6 ("above [major league] average") eventually. He ran bases now at 4 ("below average") but in the future would be at a 7 level ("very good"). The narrative paragraph section at the bottom helped explain these ranges by noting that he had "outstanding future potential," with "good raw tools" and the "desire to improve." Drake just needed training, in Lakey's view. This helped distinguish him from some other poor hitting prospect who didn't have the tools or the potential to improve.[29] It was Lakey's job to grade Drake's skills as they might eventually be.

Bodies in the present were never the actual focus of scouts: future bodies preoccupied them. That was true of the grades as well, as Hep Cronin explained: "The present grades don't mean anything. The future is what everyone cares about." That's why scouts were careful to distinguish present from future values. That's not to say that there wasn't disagreement on whether such projections could be made. Some scouting experts swore that a batter's running speed and a pitcher's velocity, for example, were fixed and shouldn't be projected. Gorman described it as "what you see is what you get," while Haak claimed that when it came to running and throwing, "a kid either has those abilities or he doesn't."[30] Coaches could teach players technique, but there wasn't really going to be much change over time.

The desire to separate future grades from present grades was also about precision, as the Phillies' 1978 manual explained in justifying the club's adoption of the bureau's distinction. "In the past, some scouts may have combined the present ability with future expectations in determining a grade for a player," the manual declared. "Now you will be able to separate the two grades and this should allow us to be more precise and refined in our grading as to a boy's ability to improve on his God-given tools."[31] Given the

```
::::::::    M. L. S. B.     FREE AGENT REPORT    10/17/74    ::::::::
-------------------------------------------------------------------------

OVERALL FUTURE POTENTIAL : 75                        REPORT NO : 1

PLAYER          : DRAKE          , KEVINMAURICE       POSITION :OF

CURRENT ADDRESS:

TELEPHONE       :                   DATE OF BIRTH :  4/12/56

HEIGHT : 6. FT.  3 IN.  WEIGHT : 200 LBS.      BATS : R     THROWS : R

PERMANENT ADDR :SAME

TEAM NAME : CABRILLO, LOMPOC,CA

SCOUT : LAKEY G          DATE :  5/27/74 RACE : N      GAMES : 4
                         INN  :   20


RATING KEY ...      NON-PITCHERS ... PRES FUT

8-OUTSTANDING       HITTING ABILITY... 2   6     HABITS......... GOOD
7-VERY GOOD         POWER............. 3   6     DEDICATION..... GOOD
6-ABOVE AVERAGE     RUNNING SPEED..... 7   8     AGILITY........ FAIR
5-AVERAGE           BASE RUNNING...... 4   7     APTITUDE....... FAIR
4-BELOW AVERAGE     ARM STRENGTH...... 6   7     PHYS. MATURITY. GOOD
3-WELL BELOW AVE    ARM ACCURACY...... 4   6     EMOT. MATURITY. GOOD
2-POOR              FIELDING.......... 5   7     MARRIED........ NO
                    RANGE............. 6   8
                    BASEBALL INSTINCT. 5   6     DATE ELIGIBLE.. 0674
                    AGGRESSIVENESS.... 6   7     PHASE.......... REG

                    PULL HITTER


PHYSICAL DESCRIPTION(INJURIES, ETC.):
=====================================
EXCL WELL PROPORTIONED FRAME WITH STRONG ARMS AND LEGS.
STRONG FACIAL FEATURES.  NO KNOWN INJURIES AND A EXCEPTIONAL
ATHLETE IN FOOTBALL.

ABILITIES:
=========
HAS BAT SPEED WITH POWER POTENTIAL. ABOVE AVERAGE ARM STRENGTH WITH
VERY GOOD CARRY.GOOD BODY, ENJOYSPLAYING THE GAME. HAS A
CHANCE TO BE A GOOD OUTFIELDER.

WEAKNESSES:
==========
GETS AWAY FROM THE PLATE SLOWLY.  NEEDS TO IMPROVE HITTING MECHANICS
AND NEEDS TO BE MORE AGGRESSIVE AT THE PLATE.  LOCKS HIMSELF UP AT THE
PLATE AND DOESN'TUSE HIS HIPS.

SUMMATION & SIGNABILITY:
=======================
HAS OUTSTANDING FUTURE POTENTIAL.  POSSESS GOOD RAW TOOLS WITH DESIRE
TO IMPROVE.  NEEDS HELP AT THE PLATE BUT OTHER TOOLS ARE THERE.  HAS
FOOTBALL GRANT FROM UCLA, WILL BE ATOUGH SIGN BUT DESIRES BASEBALL.
                         -7-
```

FIGURE 7.2: Gordon Lakey's 1974 report on Kevin Drake shows both the centrality of the present and future numerical grades and the bureau's pioneering use of computerization. Branzell Collection, BHOF

difficulty of predicting success, precision did not mean accuracy, however. Lakey's model report for the bureau misevaluated Drake entirely. Though drafted in the first round (fifteenth overall) on the basis of these optimistic projections, Drake played only four mediocre seasons in the minor leagues.

When Lakey wrote of Drake's "future potential," he was also referencing a new measure that the bureau introduced called the overall future potential. The OFP, as it quickly became known, was a number between 20 and 80 meant to combine all relevant features of the future grades. It was a single number that enabled players with a range of skills to be compared—and ultimately, of course, to be ranked for the draft.

The bureau explicitly intended the OFP to be useful for clubs' draft lists, though almost immediately after introducing it, the organization discovered it didn't work as well as intended. Scouts were simply giving grades to individual skills and then picking a number between 20 and 80 that they felt reflected their overall judgment, regardless of the individual grades they were giving. It was not, therefore, "compatible or relative to the individual numbers used in grading a boy's various skills." As a result, the bureau provided a "formula" to help scouts compute this number, a way of moving from the individual grades to a single OFP.[32] Scouts were instructed to take the grades given in eight "future" categories for position players (hitting ability, hitting power, running speed, base running, arm strength, arm accuracy, fielding ability, and fielding range, all marked on a 2–8 scale), find the average, and multiply that by 10. The resulting number would be between 20 and 80 and would be the "basic" OFP. Then scouts could take into account the remaining grades (for baseball instinct and aggressiveness) and adjust the OFP in either direction. The manual concluded by giving example OFP calculations.[33]

Recognizing that scouts would not know how to interpret the OFP at first, the manual also provided methods of connecting the metric to traditional ways of measuring prospects' potential. The OFP might be linked to value—an OFP of 75 was equivalent to a bonus of $150,000, while one of 40 was equal to a bonus of $6,000—or to prospect class—70 was a grade for a major league "star," whereas 35 was for a player with only minor league potential.[34] Such charts portrayed the OFP as a numerical equivalent of previous judgments.

Individual clubs might obviously have preferred slightly different calculations and standards, and indeed, the Texas club devised its own "standard formula for judgment" of the OFP in the materials provided to scouts in 1975. Branzell had to spell out the meaning of OFP as "overall future potential" in his manual, a sign of the calculation's novelty. In fact, the measure wasn't obvious even to experienced scouts like Branzell, who had to repeatedly scratch out and tweak the calculation guidelines in his copy. The Rangers' scouting reports also reflected the team's particular preferences: the "future" grade for running speed was blacked out, suggesting that the club believed prospects' speed would not change over time—what you saw was what you would get—and instinct and aggressiveness were moved away from the skill grades to the part of the form with the rest of the nonphysical categories. (The Rangers' reports also had a place for "socioeconomic status," perhaps providing an early indication that some players would have separate difficulties.) The New York Yankees blacked out "present" grades for hitting ability and power, as well as "future" grades for running, indicating that the team thought some grades were only relevant when they concerned the future, while others were only relevant to the present.[35] Nevertheless, nothing in any of these forms diverged too far from the basic bureau report.

The Rangers' manual admitted that despite providing a "standard formula for judgment," "in dealing with human beings, there are absolutely no mathematical formulas that will always get the job done right."[36] There were guidelines, however, that provided a way to take a collection of judgments, label them with numerical grades, and combine them into a single number. As Pries had emphasized in designing the bureau, "efficiency" in both making and communicating judgments was always a central goal.[37]

The calculation of the OFP may have provided a mechanical way to move from individual judgments about skills to a single judgment expressed numerically, but it certainly wasn't an attempt to remove scouts' discretion—on the contrary, in 1975, the Texas Rangers allowed Branzell to make adjustments to the OFP "formula" of up to 5 points, and more if he justified his decision on the report. A decade later, changes of up to 10 points could be made easily as long as the reason was noted, allowing scouts to "use their instincts." Bigger changes—say, for a "superstar" or "franchise saver" or a player with a "certain flair" that would draw fans—could be made too, but with caveats. Prospects must already have received a 55 OFP, and the total score could not exceed 80. Plus, it was "*absolutely* essential" to "explain the reason" for such a large change on the report. In short, scouts' judgments of factors beyond those explicitly specified for the OFP could intrude on their calculations but had to be clearly notated.[38]

The Rangers and Yankees were certainly not alone in adopting the bureau's forms, along with its OFP and related measures. Through the late 1970s, and especially after membership was made mandatory in the early 1980s, every club had an incentive to make its reports align with those of the bureau. That's not to say subjective judgment was eliminated. Each scout was still asked to offer his own assessment of a prospect's ability. For most clubs, though, every judgment had to be expressed using the bureau's form and scale.

Two scouts whose careers bridged this transition make visible the impact the bureau had on individual clubs' scouting reports. Joe Stephenson's four decades of scouting began in the 1950s, with notes jotted on cards about prospects' abilities (see figure 5.2). The same basic reporting system continued to be used by Boston through the 1960s. Then, in 1969, likely spurred on by the draft, Stephenson's reports for the Red Sox changed substantially. They now had grades (X [excellent], G [good], F [fair], and P [poor]) for each skill, as well as an overall mark for draft selection.[39] There were still spaces for "additional comments," but they were just

MAJOR LEAGUE SCOUTING BUREAU
FREE AGENT REPORT

Report No. 2

Overall future potential 55

PLAYER SCIOSCIA | MICHAEL | LORRI
Last name | First name | Middle name

Position C

Current Address: MORTON | PENNA. | 19070
City | State | Zip Code

Date of Birth 11/27/58 Ht. 6-2 Wt. 200 Bats L Throws R

Telephone (Area Code)

Permanent address (If different from above) SAME

Team Name SPRINGFIELD H.S. City SPRINGFIELD State PENNA.

Scout BRAD KOHLER Date 5/24/76 Race W Games 2 Innings 14

RATING KEY	NON-PITCHERS	Pres.	Fut.	PITCHERS	Pres.	Fut.	USE WORD DESCRIPTION	
8—Outstanding	Hitting Ability	4	5	Fast Ball			Habits	GOOD
7—Very Good	Power	3	5	Curve			Dedication	GOOD
6—Above Average	Running Speed	3	3	Control			Agility	GOOD
5—Average	Base Running	3	4	Change of Pace			Aptitude	GOOD
4—Below Average	Arm Strength	5	6	Slider			Phys. Maturity	EXL.
3—Well Below Average	Arm Accuracy	4	5	Knuckle Ball			Emot. Maturity	GOOD
2—Poor	Fielding	4	5	Other			Married	NO
	Range	3	5	Poise				
Use One Grade	Baseball Instinct	4	5	Baseball Instinct			Date eligible	
Grade On Major	Aggressiveness	5	6	Aggressiveness				06/76
League Standards	Pull Str. Away			Arm Action				
Not Amateur	XX			Delivery			Phase	REGULAR

Opp. Field

Physical Description (Injuries, Glasses, etc.) Graduation 06/09/76 RANGEY WELL BUILT KID.

LONG ARMS AND SOILD LEGS. MOVES AROUND BEHIND PLATE WELL FOR A KID HIS SIZE, HAS NO

KNOWN INJURIES OR CONTACTS. BUILT ALONG THE LINES OF J.C. MARTAIN.

Abilities YOUTH PLUS ARM STRENGTH AND QUICK BAT WITH CONTACT, AGGRESSIVE AND HAS

QUICK SOFT HANDS. SHOWS LINE DRIVE PWR. HITS TO ALL FIELDS, HANDLES HIMSELF WELL.

CHALLANGES PITCHER WHEN HITTING. LOVES TO PLAY. KEEPS IMPROVING.

Weaknesses LACKS RUNNING SPEED, ALSO IS A BLINKER ON CATCHING, BUT NOT TO THE

POINT THAT IT IS TOTAL HOPELESS TO CORRECT. WILL NOT BE AVG. BASE RUNNER, BUT WITH

HIS INSTINCT AND AGGRESSIVE PLAY CAN GET BY.

Summation and Signability Worth $26,000 Will He Sign YES Will Sign For $25,000 = CSP.

HAS SEVERAL COLLEGE OFFERS, A FINE STUDENT IN CLASS ROOM, HAS SEVERAL MAJOR

LEAGUE TOOLS GONE FOR HIM, BAT SPEED AND ABILTY TO MAKE CONTACT, YOUTH AND ARM

STRENGTH PLUS GOOD HANDS.

FIGURE 7.3: Major League Scouting Bureau standards, as seen in Brad Kohler's 1976 report on Mike Scioscia, were widely adopted, particularly the 2–8 scale for skills and the overall future potential calculation. Scouting Report Collection, BHOF

that—additional. The report now required Stephenson to put his judgments in a format that could be easily compared with those of other scouts.

Stephenson used this form through 1975, though by the end of that period, he was occasionally crossing out the letter scale and using numbers instead. Then, in 1975, just as the new bureau was expanding, the Red Sox adopted a modified version of the bureau's form that included full numerical grades, a distinction between present and future grades, an OFP calculation, and only brief narrative sections at the bottom.[40] This form, with only minor variations, was used for the rest of Stephenson's career, into the 1990s.

Likewise, Tom Ferrick, a former major league pitcher and pitching coach, scouted for the Kansas City Royals for three decades. By 1966, Kansas City had adopted a check-box system for scouting reports, with grades of 1 to 4 given for various skills. As early as 1967, Ferrick's forms had sections to grade present and future skills (1 being "outstanding" and 5 "poor") and small narrative portions for commentary. This hybrid form remained in use into the 1970s, though there were variations, with the back sometimes reserved for comments and the present and future grades occasionally collapsed into one grade. By 1975, like Boston, Kansas City had adopted an only slightly modified form of the bureau's report, with a 2–8 grading scale for present and future skills and a resulting OFP, and never looked back.[41]

In both cases, scouts experienced a transition to quantified reports within a decade after the draft—though clubs implemented this change at different speeds—and then, within a couple years after the formation of the scouting bureau, a transition to the bureau form or a slight modification of it. There were still holdouts. Among others, Minnesota and Cincinnati used letter grades (X for "excellent," G for "good," etc.) into the late 1970s. Other clubs, especially those that prided themselves on their own scouting operations, did not rush to adopt the OFP or the 2–8 grading scale—the Chicago Cubs' report on Biggio in 1987, for example, persisted in using a 1–10 scale for present and future values and a dollar value instead of an overall future potential.[42]

Some individual scouts rejected the move to numerical precision (and had enough standing with the front office to get away with it). Jocko Collins complained in the early 1980s that "these younger scouts like numbers, but they wind up splittin' hairs: 47 or 48, what the hell's the difference?" More commonly, scouts pushed back on the ability to isolate any given skill, taking it out of context or treating other factors as equal. Gib Bodet—not generally opposed to precision—nevertheless admitted that "the bottom line to most scouting is that all things are never equal."[43]

With such high stakes, precision provided confidence to other scouts. As Julian Mock, a longtime scout and scouting director for Cincinnati, claimed, precision was most important when it came to quantifying the probability of signing a player:

> I don't want a scout to fill out the boxes that say "Good, fair, poor" on this one. No-siree. I want that scout to tell me the percent involved. Let me know what kind of chance I'd be taking to use one of our draft picks on this kid. If I lose this kid, there's 29 other teams each picking a player before I get another draft pick and that's a lot of other good players that could be gone because of a misjudgment on this kid.

The veteran scout Bodet remembered one of his mentors, Bob Zuk, was also keen on standardization. Saying "Boy, this guy can really run" wasn't useful for Zuk. Instead, all scouts needed to use "the same language." And that language, increasingly, was numerical. So, if a scout wrote that a player could "really run," his recorded speed better be above average. In this way, front offices standardized expressions and grades, forcing scouts to "use the same terminology." Bodet appreciated that emphasis once he was himself a supervisor. "The perfect scenario," he concluded, "is to have all your scouts using the same terms in their reports."[44] A lack of precision rendered reports far less useful for comparison. Good, but imprecise, judgment didn't cut it for some scouting directors.

Clubs also had to carefully police grading standards. One problem some clubs, like the Phillies, had with scouts was that they ignored the fact that grades were meant to refer to major league ability. So if a particular grade indicated average ability in the case of hitting, that meant average ability in the major leagues at the moment of evaluation. Such a level of skill was of course extraordinarily rare among nonprofessional players (by definition, many professional players don't hit that well), yet there were still "many reports" submitted giving amateurs this grade. Consequently, the Phillies' manual warned, "Some scouts are still living in the past when there were fifty different minor leagues." It was no longer possible to waste draft picks on wait-and-see players. Philadelphia's front office wanted to make it clear that the club was only interested in those players who would eventually be able to compete at the top level.[45]

Over time clubs gave more and more explicit instructions to try to standardize grades. If grades were the foundation for the OFP, and the OFP for draft lists, then the seemingly precise and complete draft ranking would be all but useless if grades had not been properly calibrated. By the early 1990s, Chicago's National League club devoted an extensive section of its scouting manual to the grading system, noting, "We need uniformity to properly credit each player his dues"—and, perhaps more to the point, to ensure the Cubs didn't waste money and draft choices. This uniformity could only come from knowing "EXACTLY what each numerical grade and word description means" across the entire organization. Noting that some scouts graded "higher" and others more "conservative," the club urged scouts nevertheless to "endeavor to close the gap" as much as possible.[46]

For the Cubs, this meant precise tables of correspondence. Right on the form were guidelines for grading pitchers' velocity as well as hitters' power and ability. The club conveniently provided templates for people who met these guidelines (Minnesota's Kirby Puckett was a 70 hitter; Oakland's Mark McGwire had 80 power; Atlanta's Greg Maddux had a 50 fastball). Moreover, every grade implied specific values. A grade of 70 for "very good" velocity corresponded to 90 to 92 miles per hour; 50, or "average," power corresponded to 10 to 15 home runs at the major league level; and 30, or "well below average," hitting corresponded to a major league batting average of .210 to .229. Similarly, fielding grades were calibrated to major league fielding percentages, so .980 for catchers was average and equivalent to a 50 OFP, whereas for infielders .965 corresponded to the same grade. A right-handed batter who took 4.4 seconds to reach first after he made contact would receive no higher than a 4 for speed.[47] Numerical grades enabled clubs to encourage precision to whatever degree desirable.

———

In the 1960s, the keyword for scouting was "judgment." Baltimore scout Frank McGowan's article on the nature of the job, distributed to Branzell as part of his scouting manual, was titled "Judgment—The First Ingredient in Scouting." McGowan acknowledged any such judgment operated within a complex "organizational procedure," with a group of scouts and front office personnel "coordinating as a team in the judgment and acquisition of player talent."[48] McGowan's view was typical: good scouting was about good judgment.

After the draft and the establishment of the Major League Scouting Bureau, however, *objectivity* was increasingly emphasized. The Texas Rangers' 1975 manual explained that among the qualifications for a scout, above good judgment and courage of conviction, even above administrative ability and knowledge of territory, was objectivity. Just a few years later, Philadelphia emphasized that judgment remained important but defined good judgment by warning against a scout becoming too "emotionally involved with a boy and his family" and losing his "objectivity." Here objectivity was about a kind of distance, an unemotional and clear-eyed look at things as they were. By 1990, Texas had updated its manual to tell Branzell and his colleagues that scouting was now just as much about "gathering of information" as about judgment: "You must be a *digger* in gathering the right information and a *digger* in making certain it is accurate."[49] Scouts were increasingly understood to be hunters of data, recorders of data, and compilers of data.

The language of fact-finding and of emotional distancing, alongside the need to write things down, is a familiar marker of knowledge-making. Paperwork, especially the calculation of the OFP, is essential to the modern practice of scouting. Tellingly, the Yankees provided a long list of items that each scout needed to have after the draft, which were mainly related to paperwork: stationery, memo paper, envelopes, expense forms, health questionnaires, publicity forms, index cards, report forms, summary sheets, and itineraries. Scouts were instructed to always submit formal written forms, rather than calling or sending a separate memo—it was important to keep everything standardized. The Phillies made this approach clear in 1978:

> Paperwork is a necessary evil in modern day baseball. Contrary to the feelings of some, we try to eliminate as much paperwork as possible. There is good reason for the information and reports we do request; therefore we expect and will require full compliance. Above all, use the forms provided rather than letters or phone calls. This information concerning players must be in writing for future reference and it must be easily accessible. Aggressive scouts, who believe in their judgement [*sic*], should have no objection to putting their *own* opinions in black and white.

When the Major League Scouting Bureau informed scouts about "what scouting is," the first definitions included "discipline, organization, judgment, making decisions," and "writing, writing, writing."[50] Judgment had not been eliminated—it was always part of the list—but it had been made operational. One had to gather information for the club's use, with judgments expressed numerically and written down.

There were also new bureaucratic mechanisms by which scouting knowledge might be made more objective. Pioneered by Baltimore's Jim McLaughlin in the 1950s and Walter Shannon in the 1960s, the cross-checker was a scout who was promoted to oversee his peers in a particular region or even nationally (when he became known as a "national cross-checker"). Though cross-checkers appeared in proto-form before the 1970s, only after the draft and their use by the Major League Scouting Bureau did cross-checking become a standard scouting procedure. Reports would come in on thousands of players from around the country, and someone would have to sift through them to figure out how to turn the pile of evaluations into a single draft list. Cross-checkers would often see top prospects themselves, providing a different view for the front office on the player that acted as a check on the area scout's opinion. If disagreement on the player's value existed, the general manager or farm director might have to step in to decide whether to go with the usually more experienced, and higher-ranking, cross-checker or with the lowly area scout who almost certainly had more knowledge of individual players and their situations.[51]

Gib Bodet explained the role of the national cross-checker in an article he wrote for the Society for American Baseball Research's volume on scouting, *Can He Play?* Having been a cross-checker for 16 of his 40 years in the business, he called the system "flawed" but still declared it the best he knew. The fundamental problem remained that "scouts don't grade players in the same way"—"it's a fact of life, as old as the game itself."[52] Before the advent of the cross-checker, a front office guy might come out and look at the top few prospects for each scout, but with two or three cross-checkers on many teams, the scouting director could spread the work around to ensure more prospects received two different evaluations. The area scout had to be good enough to pick out the best players from the multitude of nonprospects, so cross-checkers could not replace their work. Rather, the system enabled the most experienced scouts to make the final decisions about top prospects. As a result, area scouts could express resentment when their judgment appeared to have been overruled—and for this, Bodet admitted some cross-checkers were known as "double-crossers."

The job of the cross-checker was not an easy one. A front office might have 4 or 40 reports with a "50" OFP indicated. These reports might be on pitchers, infielders, or outfielders, and a cross-checker had to go through and recommend a player to draft first. Moreover, not all 50 OFPs were equal, since scouts might have calculated that number in different ways. If hitting was more important than defense for a particular club, the front office also had to take that into account when looking at OFPs.[53]

Sometimes, though, the cross-checker needed to do more than simply compare equally rated prospects. He had to compare one scout's 50 to another's 45 or one scout's top prospect with another's. As Atlanta's Hep Cronin explained,

> Before the draft, you make up a final list of the kids in your territory in rank order by their OFP rankings. The scouting director and the national cross-checker get lists from all the area scouts like me. Then they combine all the rankings and come up with a master list to use in the draft. To do that, they have to know the area scouts and know how tough they grade. No two scouts are exactly alike—some are low, some are high.

Scouts might also have different ranks within an organization. The judgment of "super" scouts or "special-assignment" scouts might have more weight with the front office. Clubs had to calibrate a range of grades from a range of scouting opinions.[54]

Too much standardization, however, was also bad—especially if it encouraged uniformity. When veteran scout Bob King was asked what kind of advice he'd give young scouts, he replied, "Be honest about your own opinion and don't think that you have to have the same opinion as your scouting director or your supervisor. Quite honestly, you're more value[d] if you have a different opinion." Cross-checkers and scouting directors also didn't want scouts sharing notes or calibrating their grades with other clubs' scouts. Objective judgment did not mean uniform judgment.

Bodet didn't like scouts to even ask others about opinions; he wanted them to be independent. This reluctance was a common refrain among scouting directors and head scouts—Ray Poitevint of Milwaukee in the early 1980s similarly complained about scouts who consulted each other to know whether a grade should be "4" or "5." Scouting director Doug Melvin decried scouts who fraternized with or talked to other scouts about their views. Bodet prided himself on being a cross-checker who graded players before consulting the area scout's full report, ensuring that he was not "influenced" by a positive write-up. He could see the player objectively. Similarly, Philadelphia justified the position of the cross-checker by noting that it helped correct for scouts who freely exchanged information with competitors instead of valuing players independently. Too often, club officials complained, violation of the "closed-mouth" policy by "loose tongues" not only helped competitors but also hurt the club, especially in the resale market for players Philadelphia was trying to unload. Other scouts were "enemies and are out to beat you in any way possible," the club manual declared.[55] Scouts were paid to relay information to their own club, not to share it with the competition.

There were some good reasons for coordination from the perspective of an area scout. For one thing, it taught new scouts what a "4" fastball or arm strength looked like. For another, it ensured that one's neck wasn't too far exposed—if all the clubs were overrating a player, and you did too, at least you just looked like one of the bunch. If you were alone in overvaluing a prospect, however, your club was potentially out a great deal of money, and you would look incompetent.

———

Though their central job was to assist in the numerical ranking of prospects, scouts remained prized for their knowledge of people. Bodet himself admitted that it took a "lot of skills" to talk with prospects and their families about money and bonuses "without being misunderstood." And he was also clear that scouts needed good working relationships with local high school and college coaches so that they could be told about prospects they might have otherwise missed. Scouts needed to deal delicately with coaches who invited them to view a player who turned out to be a nonprospect, or with parents, so as not to tip the club's hand about a player's real worth and undermine negotiations. Stewart echoed Bodet in saying that even in the age of the draft, when it was not clear which team would draft which prospect, the personal relationship remained essential because scouts still had to sign drafted players: "The quicker you get to know a prospect, the more he's going to remember you, the more trust you're going to build up with the family."[56] Scouts had a dual set of tasks: hard grading of talent and soft negotiating with families and agents once a player was drafted. This was especially true for amateurs drafted after the first few rounds (i.e., the vast majority of them), when scouts and clubs were trying to convince prospects to take a small bonus and possibly forgo college or more lucrative work in order to have a chance of playing in the major leagues. Scouts may focus their own memoirs on the players who made it big, but by necessity most of their time was spent grading and negotiating with the great mass of players drafted in the later rounds who would never make it to the majors.

Scouts were known for their networks of people, their communities of correspondents pointing them to possible prospects otherwise hidden away. Scouts have long maintained extensive networks of volunteers, part-time "bird-dogs," and others. Gene Bennett, a Cincinnati scout for 58 years, maintained at least 40 part-timers under him. A system like this required a great deal of trust and generated its own economy. Favors, gift exchanges, and tip-offs were the currency of such relationships. Bennett worried constantly about cultivating his network, lest his informants offer their information to another scout. Clubs themselves pushed scouts to be careful in creating these networks. The Phillies were perhaps more explicit than most when they admonished their scouts to hire "competent part-time help and contact men. Do not

hire an old buddy looking for an easy job." Scouts needed people
with "reliable judgment" and loyalty to ensure quality information
about prospects.[57]

More importantly, general managers and scouting directors
put a great deal of emphasis on knowing their scouts. Cincinnati's
Julian Mock ran a three-day training session every year for the
club's scouts, ensuring each knew how to communicate judgments
effectively. The Major League Scouting Bureau taught its scouts
consistency through tests and quizzes. Similarly, longtime scout
and scouting director Fred McAlister explained that he only hired
scouts once he knew he could trust their judgment:

> It was always an individual process and by that I mean we didn't
> give a scouting test and hire the guy with the best score. I talked
> with the guy to find out how much he knows about baseball.
> I liked to talk to them a lot about baseball and how to look at
> players to reinforce the fact that they shouldn't be afraid of
> making a mistake.

He would thus take prospective scouts to games and then ask them
afterwards how particular players did.[58] The point was to figure
out whose judgment was trustworthy.

General managers and scouting directors understood this
need for human knowledge. Kansas City's Stewart described it as
"scouting the scout." Each scout was different, and it was crucial
that those differences be known in the front office: "As much as
we're all objective, and try to look at each player the same, we
all have our favorites and our quirks." Stewart spent time with
scouts, riding around with them on trips to get to know them
and their families, whether they were optimistic or saw potential
injuries around every corner. He also kept a "sheet" on scouts he
employed, which provided an "old-fashioned scale" for him to
evaluate every one.[59] This system obviously aimed to ensure that
a scout's production exceeded the money the club paid him, but it
also sought to learn how a scout's individual judgment translated
into his grades.

Baltimore's scouting director Jim McLaughlin similarly warned that trying to read a scouting report without knowing the person behind it was risky. "The scouting director better have a great memory of his own," he claimed, "because he's gonna get most of the good stuff in phone calls. A scout calls up, and maybe he's excited, and he gives his whole sense of a player. A few days later, his report comes in the mail, and the player sounds like John Doe on paper." There was also the problem of scouts differing in grading. Some would "use grades inconsistently," he noted. "So on the subject of written reports I'm a little cynical."[60] McLaughlin's cynicism wasn't based on the fact that the ability to throw a fastball couldn't be graded: it was that a scouting director or general manager had to know something about the scout in order to make sense of his report.

Lou Gorman, who served as a farm director, director of player development, and general manager with a variety of teams, also believed that in order to evaluate prospects, scouting directors had to know their own scouts—that's why when new ownership or management came in, they brought their own people with them. Management, after all, had no idea whether to trust unknown scouts' judgment. Human knowledge was always essential: "Generally for your cross-checkers and your advance scouts, you use your top guys and you use them all the time," said Gorman. "When you come here, though, you don't know them. You don't know their judgment. How conservative or liberal they are. You learn it as you work with them." What was a "5" to one scout was a "4" to another, and it might take a while for a scouting director or general manager to learn that, but such information was "vitally important," because the front office could "misread" scouts. After a club had drafted a player on the basis of some written report, a scout might say, "Well, I didn't really say that." Regardless of what the report indicated, it was important to know the scout, Gorman concluded: "The more you're familiar with the people, the more that you work with them, the more secure you feel in dealing with them." Clubs could get "burned" if scouting directors didn't know their scouts well enough.[61]

Gorman evaluated his staff, but he also met with them as a group five or six days a year after the season was over. They'd watch games and discuss kids together,

> have a cocktail party and get to build a little morale, get to know them better. We do that every year. You're always constantly working with them. You'll call them on the phone. You'll have them call you on the phone a lot. . . . From their standpoint, they get to see the whole system, to hear other scouts evaluate. You do that constantly. It's the way you build morale, it's the way you build judgment, and it's the way you build an organization.

Judgment—organizational, bureaucratic judgment—didn't come from grade standardization alone, or from well-worded scouting manuals. It also came from talking and learning about colleagues, from checking in. It was ultimately "built." To Gorman, even if scouting "will never be an exact science," if "you've got people with good judgment and you know their judgment is good, you can push your neck out for them because most of the time, they're going to be right for you."[62] For Gorman, objectivity and reliable scouting data emerged not from turning scouts into machines but from treating them like colleagues and friends, building trust and morale.

This level of familiarity was one reason Major League Scouting Bureau reports were read differently than those of one's own club. Ellis Clary found the bureau scouts' grading unreliable, in part because they didn't report to any particular general manager. "They find players who maybe can play the game, and they put numbers on them kids that's much more than they really are," he said. "The numbers are just ridiculous, and I don't even pay attention to them." The bureau was aware of such sentiments and eventually tried to ensure that top prospects were vetted by multiple members of the bureau, known as a "three man evaluation" team, comprising the scout, the supervisor, and the national cross-checker. After San Francisco joined the bureau, farm director Jack Schwartz was initially worried about firing most of his 40 scouts since it had taken three decades to build up his scouting operation. But, he later noted, "I've gotten to know the bureau scouts better, and I think I'll be able to rely on their judgment in the future. I know now there are some scouts I can trust as much as ours."[63]

Still, bureau scouts weren't usually known to general managers the way a club's own scouts were. And bureau scouts were not writing for any particular general manager—they were meant to produce decontextualized facts. As a result, some general managers concluded that bureau reports were overly noncommittal, with nothing at stake in their evaluations. Indeed, Joe Brown, one of the main pushers of pooled scouting, worried that after the creation of the bureau even he would have to "learn how to judge the reports," because scouts might be conservative at first, knowing that many teams would be reading their reports and not wanting to promote a bust. Furthermore, these reports would, as always, require double-checking on the parts of individual clubs. Bob Nieman, a longtime scout for Cleveland, sounded a similar warning. He worried that the inevitable variation in reports would require a supervisor who knew the scouts—"It's as important for a scouting supervisor to understand his scouts and know how to evaluate their reports as it is for a scout to recognize talent." The name on the report mattered a lot to clubs: when Cubs scout Lennie Merullo moved to the Major League Scouting Bureau in the 1970s, he recalled that the club thought his relocation was a huge benefit—after all, they still got his judgments on players through the bureau reports they received, but now they didn't have to pay his salary.[64]

Howie Haak evocatively described most bureau reports as "faceless," because "our scouting director doesn't have a feel for the grades or the habits of whoever wrote the report, the way he would for someone on his own staff." This idea of the "faceless" report is essential to understanding the interplay of subjectivity and objectivity in scouting. Reports were designed to calibrate judgment, calculate an OFP, and ultimately create a robust and reliable ordered list of prospects for the draft. But there was never an illusion that a report's precise numbers made its author irrelevant.

Rather, the "face" of the report—its author—remained important to general managers, even if not always important enough to justify an extensive scouting staff in addition to the bureau. That's why the bureau eventually instituted an annual predraft meeting with each club that desired it, during which the scouting directors could sit down with bureau scouting supervisors and cross-checkers to confidentially discuss the reports on particular players. The bureau also held a general discussion of its list of top 800 prospects, during which scouts had a chance to explain their reports in person. The bureau was aware that scouting judgments were never "views from nowhere." They were *always* someone's judgment. Each scout had idiosyncratic tastes and preferences, and their judgments remained inescapably subjective. But their views could still be made into reliable knowledge.[65]

————

Joe Branzell witnessed his profession change dramatically between the early 1960s and his death over 30 years later. Scouting had long been a part of the business of baseball, but increasingly it also became bureaucratic, with standardized forms, numerical scales, and cross-checkers performing quality control. What was once a hunt for players turned into a process of evaluating and measuring. Nothing exemplified this transformation more than the invention of the overall future potential metric.

The calculation of the OFP has remained a central tool for major league clubs and scouts. That's to be expected. The practice of labeling complex phenomena with numbers is a way of life in the late modern world. Credit scores for job applications, blood pressure guidelines for hypertension diagnoses, public opinion surveys—the reduction of phenomena to a single number is said to be scientific, rational, and objective.[66] In many regards, the calculation of an OFP works similarly, though without any strong links to academic research or formal statistics; rather, its emergence was an internal, bureaucratic development within scouting.

Moreover, unlike some examples in the history of quantification, the OFP was not intended to replace personal knowledge. Scouting reports were *always* signed and *never* impersonal—scouting directors cared deeply about a report's author.[67] The OFP and similar bureaucratic measures arose not as a ploy for scouts to gain legitimacy or for their methods to obtain the patina of scientific rigor, but as a way for clubs to compare individual judgments about thousands of prospects and emerge with a ranked list suitable for navigating the draft.

New scouting reports and metrics were just one of the technologies the Major League Scouting Bureau promoted to help make scouting more reliable. From the start, the bureau made technology a selling point. Reports were standardized, compiled electronically, and distributed daily through an electronic network from a central computer. Later, the bureau would be among the first to compile a video library of prospects that clubs could access in order to scout players without needing to see them in person. By the early 1990s, the bureau was filming nearly 300 top prospects every year and disseminating the video to all clubs. Tellingly, when the bureau downsized dramatically in 2014, reducing its ranks from 60-something scouts to fewer than two dozen—it retained its administrative emphasis, continuing to provide identification of younger players, as well as medical information on and video coverage of possible draft picks.[68] As its name has always implied, it is first and foremost an information bureau.

The Major League Scouting Bureau was central to establishing the modern practices of scouting, from cross-checking to report compilation, computerization to OFP calculations. Several bureau staff eventually moved into executive positions with clubs, which only increased the organization's influence. The bureau also played a central role in training scouts across the major leagues through its Scout Development Program. For a quarter-century after Pries created it in 1989, this so-called "Scout School" was a nearly two-week program in which potential scouts were introduced to the terminology, grades, standards, and techniques used by bureau scouts. (Pries had taken over the bureau after Wilson retired in 1985.) Attendees were sponsored by a club or the bureau itself— that is, they were candidates to become professional scouts, as opposed to interested laymen. Scout School days involved two hours of class followed by a session devoted to watching a game and then a debriefing session discussing reports about assigned prospects. The point of such sessions was to provide uniformity across the league in the production of the "formula for judgment" about players. The school provided a binder not only explaining the grading scheme and "what to look for," but also suggesting standardized language for reporting. Something like base running was not just a matter of "quick" or "slow"; instead, it could be characterized using over two dozen recommended phrases, including "clogs bases," "green-light guy," "smart on bases," and "steady on bases."[69] Scouts needed to be taught not just to recognize future talent but also to use precise language to describe it. Whereas before such training might have been offered club by club, now the bureau provided it to the prospective scouts of all clubs. This centralization made sense, given the advantages of having club reports match bureau reports. Even if individuals and clubs diverged on their judgments of specific players, those judgments could still be calibrated and standardized.

One additional advantage of the increasing precision of reports was that they enabled ever-finer evaluations of scouts themselves. After 1965, every general manager knew that a scout could not simply be judged by the players he signed, since so few of those he saw would be drafted by his club. Moreover, since the vast majority of prospects would fail, simply saying "no prospect" in every case meant that a scout would predict correctly a great deal of the time. Clubs wanted a better way of evaluating scouts, and one solution was to use the scouts' own ratings. Philadelphia made this approach explicit back in the 1970s:

A scout is judged on the players that he recommends that we do sign and those that he recommends that we are not able to draft or sign, as well as players selected or signed by other organizations that he has not recommended. Such records are maintained by the Philadelphia Office on each scout. This rule and the future development of such players really places the emphasis on evaluation and a scout's judgment.

Into the 1990s, the Cubs continued to demand almost laughably specific predictions from their scouts—not just ratings on future skills for each prospect, but also a "comfort zone" rating of their confidence, an "impact" rating from "definite all star" down to "no prospect," and a "major league role" prediction as to whether the player might be a fifth outfielder or a number-two starting pitcher.[70] Such evaluations were obviously not about players—the process of becoming a number-two pitcher could never be predicted with any accuracy—but rather about scouts. Clubs wanted to be able to go back and see what kinds of ratings the scouts had given to what sorts of players and if some scouts were more precise or prescient than others, or if there were systematic biases not even apparent to scouts themselves (perhaps they overvalued relief pitchers but undervalued left-handed outfielders).

The changes that occurred in the practice of scouting over the twentieth century resulted in the objectification of players—turning them into numbers—even as it respected the expertise of scouts. The numerical evaluations of players could be turned back on scouts, however, enabling clubs to quantify scouts' own predictive ability. Scouts too were judged by the numbers.

Not surprisingly, the requirement for precision rankled some scouts. Stewart admitted that whatever effort goes into calibration, "no scouting scale can take away all the ambiguity, of course. We are fallible human beings judging other fallible human beings, and not only that, you're judging them at 17 years old based on what you think they'll be at 25." Bodet, despite his efforts at calibration, ultimately sounded a similar note: "Scouting isn't a science at all. What we're doing is establishing who the kids are and what their potential may be."[71]

Bodet may have been too modest. It is precisely in the modes, methods, and tools of establishing prospects' potential that the history of scouting parallels changes in other disciplines devoted to data analysis. Scouting may never be a "science" in the way astronomy or physics is a science, because, as longtime general manager Pat Gillick admits, "it's not a very exacting science. No matter how much testing you try to do, you never really know how a young guy's going to react to a pro situation until he plays in one."[72] Gillick and Bodet are right—scouting isn't much of a science, if by "science" they mean having universal laws and perfectly replicable findings. But most sciences don't have universal laws or perfectly replicable results. Rather, scouting is a science in that it serves as a well-developed and well-crafted set of heuristics for arriving at stable, generalizable, and reliable facts about the natural world. Like many scientific practices, scouting is increasingly centered on the use of ever-larger datasets and dependent both on the technologies required to collect the data and on the expertise needed to manage and analyze them. It does not yet involve "big data" or algorithmic processes, but, like all data-driven sciences, scouting relies on the search for meaningful patterns in the data, induction from existing data, and continual adjustment of inferences on the basis of new data.[73]

Scouts envision themselves as talent hunters, rugged individuals on their own in the search for diamonds and ivory. But after the draft was initiated, a longstanding truth was made painfully evident: scouts are cogs in a giant bureaucratic machine. Veteran scouts revealed this by the way they complained about their work after the draft. Ed Katalinas said that "scouting's more of a numbers game now. These young guys won't have the chance to gamble like that or to romance a prospect the old way. They've become graders, and they can't afford to fall in love with talent." Bobby Mattick thought that after the draft clubs had become "too scientific" about scouting. McAlister said that after the draft clubs couldn't promise anything to the player: "It had become a matter of numbers."[74]

When Biggio was eligible to be drafted, scouts were looking to put a number on him. To do so, they had to participate in an objectivity machine that took in written reports from multiple people who had watched him play and manufactured an overall future potential figure, a draft order, and a corresponding price tag. These judgments were, of course, contested and contestable. But the need to express Biggio's potential in numerical terms was not.

Scouting in professional sports may remain one of the most financially consequential subjective judgments in the modern world: one fallible human evaluating another, with millions of dollars at stake. Nevertheless, to make strong or clear distinctions between scouting and rigorous, objective knowledge making is a mistake. The line has always been far too blurred. Scouts themselves were complaining that their work had become "too scientific," too much a "matter of numbers" by the 1970s. They were not referring to the influx of sabermetricians or to advanced analytics. They were referring to the complex set of practices by which they were trained to see a prospect play a game, grade him using specific numbers, and calculate a single representative value for the future potential of that person, a value that then was used to make a single, ordered list of prospects throughout the country. They were referring to the transformation of subjective judgments into objective knowledge.

Conclusion

The history of scouting and scoring suggests that the human sciences are increasingly data sciences. We now commonly understand and evaluate each other and ourselves through numerical data. In turn, the data sciences are inescapably human sciences. Data are not only often *about* people, but also produced and interpreted by people.[1]

Neither scouting nor scoring seems like textbook science, of course, and most baseball people are quick to distance themselves from the perceived pretensions of academic knowledge. Scorers and scouts are nevertheless concerned with creating reliable, usually quantified, data about a player's value, knowledge backed by bureaucratic institutions, disciplinary expertise, and formal and informal peer-review networks. This infrastructure supports the claims they make by facilitating the collection, verification, and communication of their assessments.

Baseball is just one of many sports for which proponents of data, quantification, and mathematical methods propose to fundamentally replace expensive, messy, traditional forms of human expertise with cheap, automated, cutting-edge technologies and analytics. The same, or a parallel, story could also be told about professional hockey, football, or basketball.[2] Baseball is distinct in a few key ways, however, especially in its sizable gap between nonprofessional and professional competition, which makes the amateur scouting process more fraught. Fewer than 1 out of every 10 baseball players drafted ever plays a day in the major leagues. And it isn't just that the low draft picks skew the statistic: of the 506 first-round picks in the first two decades of the draft, only 320 reached the major leagues, a 63 percent success rate. That means that of the top three dozen or so amateur players in the country each year in that period, only slightly more than half ever played a day in the major leagues. (More recently the numbers have improved slightly: from 1996 to 2011 about two-thirds of first-round and supplemental round picks played at the top level.) In football and basketball, nearly all first-round picks play at the top level at least for a short time. Unlike these other sports, the number of amateur baseball players immediately ready to play at the major league level is so small as to be laughable: in the first 25 years of the draft, only 17 players went directly to the major leagues, and only 4 of those never had to go back to the minors for more "seasoning."[3]

Perhaps more importantly, baseball's use of data has been far more extensive and occurred far earlier than in other sports. Though baseball's data were hardly "big" even into the early twenty-first century, with the advent of Statcast, data collection systems are producing huge amounts of video, movement, and tracking data in addition to thousands of statistical measures.

Because baseball has become synonymous with the promises of data analytics and the power of replacing traditional forms of expertise with new, data-driven ones, it also provides a way to recast our understanding of the meaning of data in the modern world. There are many ways to criticize the strongest claims of data scientists, and increasingly critics have begun to push back on some of the field's more expansive claims.[4] My own approach has instead been to look at what is understood as a widely successful application of data analytics—baseball—and ask what exactly counts as data in that setting.

The history of scoring and scouting certainly undermines any clear distinction between the claimed objectivity of scorers and the subjectivity of scouts. Both attempt to make assessments of value standard, reliable, and useful, but neither group has found ways around human judgment altogether. Switch the scorer, and some hits will become errors in statistical summaries; switch the scout, and some 4s will become 3s on reports. General managers and scouting directors responsible for interpreting those numbers need to know the scouts personally—their tendencies and habits, their desires and faults—in order to make sense of those reports. Official scorers come together every year to try to standardize calls, while an elaborate review system enables questions of judgment to be appealed to a higher authority. The case of baseball is different from other situations in which numbers make knowledge impersonal. It continues to make a great deal of difference *who* is filling out the scouting report or statistical summary; discretion and arbitrariness may be minimized as much as possible, but subjective judgments remain essential in both scouting and scoring.[5]

Baseball also points us to the ways that changing valuations of human labor and the changing marketplace in which labor is priced have been central to increasing quantification. The trajectories of both scouting and scoring had inflection points in the 1970s, precisely the time when the labor costs of players were changing dramatically. In 1964, on the cusp of the draft, the average player earned (in 2016 dollars) roughly $100,000; in 1974, he earned $200,000; and in 1984, $750,000.[6] That is, the *average* player went from comfortable to rich in that period.

For those analyzing player statistics, this had two major implications. First, there was an ever-greater need to quantify performance through reliable statistics, because performance metrics became an integral part of many players' contracts. Second, there was increased attention paid to measuring the actual worth of a player, so that it was clear whether spending millions on a long-term contract was justified. The very rise of Bill James–style analytics in the late twentieth century is evidence for the increasing value put upon human labor.

For scouting, expensive free-agent deals meant that the draft was the only time frugal clubs could sign top players to affordable contracts. This put pressure on clubs and scouts to find and draft the highest-quality players at the lowest prices. This approach would not just guarantee a better team on the field but potentially also produce a blue-chip trading prospect who could be profitably cashed in later. Of course, there was also more pressure not to waste picks on those with problematic injuries or psychological deficits, increasing the need for clubs to find ways to measure and assess prospects' bodies and minds.

Baseball's long obsession with data also points to the ways that the claims of novelty and revolution accompanying the advent of big data must be carefully qualified. The hope that additional data, better data, or more automated data will somehow take us to the promised land of eliminating subjectivity and bias is not new. The craze for accumulating comprehensive collections of data goes back at least to the early seventeenth century—and perhaps even further back, depending on who's counting. Educated people have long been overwhelmed by avalanches of data—which seemingly become greater every year—and methods of categorizing, analyzing, and making sense of those data have been developed in turn. Card catalogs, encyclopedias, scientific indexes, digest articles, and many other technologies have come into existence to make sense of all the possible data.[7] The size and processing capabilities afforded by modern electronic computing represent fundamentally new practices, even given that computers were based explicitly on previously existing ways of organizing and processing data. The fact that databases were once copied into books by hand, then were produced with punched cards fed through a mainframe computer, and now are simply uploaded and downloaded from anywhere in the world points to a difference of degree, not of kind. MLBAM may be doing entirely new things with Statcast and other data collection efforts, but at the core of such initiatives are the techniques and code of Project Scoresheet. It might seem like human expertise was replaced with machine expertise, but that's only a consequence of the success of reformers in claiming their efforts were revolutionary.[8]

The acknowledgment of scouts' extensive collection and use of data is also a good corrective to the way the claimed "revolution" wrought by data analytics underplays the degree to which mathematical modes of thought and analysis have long been lurking in supposedly qualitative fields. Practitioners of art, literature, and history, though often superficially opposed to numerical reasoning, have extensively used mathematical concepts and insights, from Albrecht Dürer's sixteenth-century engraving *Melancholia* to James Joyce's modernist novels.[9]

Perhaps most important, thinking about how we know what we know about baseball helps recast the relevance of expert judgment. We are often told now that we live in a world where expertise is in crisis and where algorithms can replace human judgment in domains ranging from flipping burgers to making medical diagnoses. Data aggregation and analysis have played a key role in what a recent book has called "the death of expertise." Likewise, Viktor Mayer-Schönberger and Kenneth Cukier make the diminishing need for experts clear by situating baseball analytics under the heading "The Demise of the Expert" in their book on big data. As the authors explain, as "data-driven decisions" "augment or overrule human judgment," subject-matter experts' "supremacy will ebb." Though such bold claims are not new—the pioneering statistician Karl Pearson said much the same thing over a century ago—the widespread fear of experts' demise seems novel. Even scholars have begun to debate whether or not academic studies of the origins of expertise and its connection to scientific knowledge might have facilitated widespread skepticism toward experts.[10]

When datasets and the tools to manipulate them become widely available, the democratization of knowledge is indeed heralded as a way to challenge insular subject-matter expertise. Recent debates over the existence of voter fraud in American elections, for example, were based largely on inferences from small segments of national survey data at odds with the judgments of most experts. As with climate change or smoking and cancer before it, expertise can be challenged in a politically effective way simply by sowing doubt. Critics complain that the use of statistical tools as tests of truth leads to real causal mechanisms being doubted because the statistics of causal claims can be extremely complicated. On the other hand, statistics can also create doubt by supporting too many spurious claims as a result of p-hacking and data dredging.[11] Bill James warned about these problems as they applied to baseball data back in 1982, noting that "no sabermetrician has ever discovered anything of interest by compiling large stacks of numbers and shaking them vigorously to see what happens to fall out."[12] For James, mathematical analysis began not with numbers but with good questions—questions that only those well versed in baseball would know to ask. Expertise hasn't been replaced by data; it is a precursor for knowing what should count as data and what questions could be answered by data.

Claims of declining expertise are often subsumed within narratives of insider expertise being challenged by outsiders. As Mayer-Schönberger and Cukier explain, data scientists have the virtue of being outsiders, able to see past "superstitions" and "conventional thinking" within groups because they "have the data." Michael Lewis himself describes his own *Moneyball* story as pitting outsiders and renegades devoted to data analysis against "The Club"—the insider, established powers in baseball. Though it is a leap from *Moneyball* and data science to the rechristening of falsehoods as "alternative facts" in 2017 by a special advisor to the president of the United States, self-styled outsiders, armed with a bit of data, have had success in challenging expert consensus. Indeed, Oxford Dictionaries' 2016 Word of the Year, "post-truth," was defined in terms that clearly echo the claimed scouting–scoring dichotomy: "Relating to or denoting circumstances in which objective facts are less influential in shaping public opinion than appeals to emotion and personal belief." Data analysts' downplaying of expertise has ironically made it harder for "objective facts" to triumph over "personal belief" in a world in which everybody is a putative expert.[13]

By looking at the ways scouts and scorers actually come by their information, however, such insider–outsider distinctions break down quickly. The so-called outsiders in data analysis are now very much on the inside. Bill James and Tom Tippett were eventually hired by the Boston Red Sox. John Thorn traded on his editing and publishing credentials to become the official historian of Major League Baseball. Dave Smith's Retrosheet as well as Gary Gillette's Hidden Game Sports supply historical and current play-by-play data to Baseball Reference and other media companies. Among scouts, some of the early advocates and adopters of precise, quantified metrics had long careers in management, with Gordon Lakey, Jim Fanning, Jim McLaughlin, Jim Wilson, and Art Stewart all serving as scouting directors or general managers. Neither the history of scouting, nor that of official scoring, suggests that data, or quantification in general, are mainly used by outsiders to unsettle insiders. In fact, it was the gradual expansion of bureaucracy (whether the increasing involvement of Major League Baseball's front office in scoring decisions or the development of cross-checkers and standardized scouting reports) that emphasized the importance of making judgments precise, consistent, and numerical. The history of scorers and scouts suggests that the use of numbers increased not because outsiders wanted to threaten the establishment but because establishment figures were looking to make inherently subjective judgments more reliable.

Both senses of expert judgment—that of knowing how to collect and manipulate data and that of knowing how to bring those data to bear on a field's important questions—matter. Henry Chadwick, Pete Palmer, and Cory Schwartz, as well as the hundreds of scouts who have routinely expressed judgments numerically, never thought the production of numerical data required the elimination of expert judgment or obviated the need for human labor. They knew well how much expertise and labor was required to make facts stable and credible.

For both scorers and scouts, however, the roles of expert judgment and human labor have all too often been rendered invisible. Scouts don't think of themselves as producing data at all, and their own narratives of what they do hide the extensive role they play in quantifying, calculating, and establishing numerical data points for prospects. Scorers' work is often hidden simply by the presentation of statistics in tables as free-floating numbers with no relation to the processes by which they were rendered stable and reliable. If you know where to look, though, the knowledge is hiding in plain sight.

That's certainly the case with two documents evaluating Craig Biggio. The first is from a 1987 scouting report made during Biggio's last year in college. The second is taken from the seminal book by Pete Palmer and John Thorn, *Total Baseball*, published just two years later.

Both clearly rely on numbers and attempt to use them to evaluate Biggio's performance and ascertain his value. These documents come from very different genres, however. One represents a private attempt, intended for just a few people, to convey Biggio's potential. The other is a public record of his performance in the previous year's season. Each engages in a distinct mode of evaluation.

By bringing them together, however, by thinking of them both as physical records of an attempt to make judgments about Biggio's ability, we can start to reveal the infrastructure by which data about baseball are created. In Biggio's case, that infrastructure includes not only the labor of official scorers in adjudicating hits from errors and earned from unearned runs, but also the work of data collectors and database builders like Pete Palmer in collecting and verifying statistics, as well as the choices made about the definition of statistics and which ones matter enough to be kept and displayed. It includes, too, the experience of scout Tom Ferrick, as well as his use of the basic Major League Scouting Bureau form; the way the reports were made comparable to other reports on Biggio's ability; the fact that Ferrick not only measured Biggio's running time to first base down to the tenth of a second but also

KANSAS CITY ROYALS

FREE AGENT REPORT

OFFICE USE
Report No. _____
Player No. _____

Overall Future Potential _____ 60

Nat'l. Double Check   Yes ✓ No _____

Scout's Report # ____ 1 ____   Scout ____ FERRICK ____

PLAYER ____ BIGGIO ____ CRAIG ____ Pos. ____ C ____ Date ____ APR 15 87 ____
            Last Name   First Name   Middle Initial

School or Team ____ SETON HALL UNIVERSITY ____ City and State ____ S. ORANGE ____ NJ ____

Permanent Address _____
                    Street   City   St   Zip   Phone

Current Address _____
                  Street   City   St   Zip   Phone

Date of Birth _____ Ht. 6:0 Wt. 185 Bats R Throws R DATE ELIGIBLE JUNE 87 PHASE R

Game Date(s) ____ APR 15 1987 ____ Games ____ 3 ____ Innings ____ 21 ____ Graduation ____ JUNE '88 ____

No. RATING KEY	M.P.H.	NON-PITCHERS	Pres.	Fut.	PITCHERS	Pres.	Fut.	MAKEUP				
8—Outstanding	94-	Hitting Ability	3	4	Fast Ball				Ex.	Good	Fair	Poor
7—Very Good	91-93	Power	3	3	Life of Fastball							
6—Above Average	88-90	Running Speed	7	7	Curve			Habits	4	✓	2	1
5—Average	85-87	Base Running	5	5	Control			Dedication	4	✓	2	1
4—Below Average	82-84	Arm Strength	5	5	Change of Pace			Agility	✓	3	2	1
3—Well Below Ave.	79-81	Arm Accuracy	5	6	Slider			Aptitude	4	3✓	2	1
2—Poor	0-78	Fielding	5	6	Other			Phys. Mat.	4	✓	2	1
		Range	5	6	Poise			Emot. Mat.	4	✓	2	1

USE ONE GRADE	Hitting: (✓)	Running					Baseball Inst.	4	✓	2	1
Grade On	Pull .... 3 ✓	Time To	Arm Action	EX 4	GOOD 3	FAIR 2	POOR 1	Aggressive-	4	✓	2	1
Major League	St. Away .... 2 ____	1st Base		3/4	OH	SIDE	OTHER	ness				
Standard	Opp. Field.. 1 ____	4.0 4.1	Delivery	4	3	2	1	OVERALL	4	✓	2	1
		Gun Reading ____ to ____ MPH										

Physical Description (Injuries, Glasses, etc.) WIRM - LIVE BODY. NO GLASSES. NO KNOWN
INJURIES. WELL PROPORTIONED

Abilities AGGRESSIVE - TAKE CHARGE TYPE. NOT AFRAID OF CONTACT AT
PLATE. QUICK - AGILE IN ALL MOVEMENTS. HANDS - RECEIVING - BLOCKS
BALL IN DIRT WELL. ARM - SML - CLOSE TO 6 ML. QUICK RELEASE, ALERT
IN GAME - HUSTLES WELL - FAIR CONTACT WITH BAT. LINE DRIVE TYPE HITTER

Weaknesses NONE APPARENT AT THIS TIME.

Signability: Ex. ____ Good ____ Fair ✓ Poor ____ Worth: $ 55-60000
MIKE SHEPARD - COACH - ALWAYS GETS INVOLVED IN SOME WAY
WITH SIGNING HIS PLAYERS. CAN BE TOUGH.

Makeup Evaluation and Player Summation GOOD MAKEUP - MENTALLY TOUGH. HAS THE PHYSICAL
TOOLS TO BE M.L. CATCHER. NOT A POWER HITTER. CONTACT + LINE
DRIVE TYPE. HIS BAT WILL DETERMINE NO. 1 OR 2 STATUS. CAN
ALSO STEAL BASES. GOOD SPEED FOR CATCHER.

FIGURE C.1: Tom Ferrick's 1987 report on Craig Biggio. Hart, *Scouting Reports* (1995), p. 14

BIGBEE-BINKS

Player Register

■ CARSON BIGBEE

Bigbee, Carson Lee "Skeeter" b: 3/31/1895, Waterloo, Ore. d: 10/17/64, Portland, Ore. BL/TR, 5'9", 157 lbs. Deb: 8/25/16

YEAR	TM/L	G	AB	R	H	2B	3B	HR	RBI	BB	SO	AVG	OBP	SLG	PRO	/A	BR	/A	PF	CHI	RC	TA	SB	CS	SBR
1916	Pit-N	43	164	17	41	3	6	0	3	7	14	.250	.285	.341	.626	88	-2	-3	105	22	19	.585	8		0
1917	Pit-N	133	469	46	112	11	6	1	21	37	16	.239	.301	.288	.589	81	-10	-9	100	61	46	.549	19	15	-6
1918	Pit-N	92	310	47	79	11	3	0	19	42	10	.255	.344	.319	.663	97	2	0	106	75	40	.693	19	20	-2
1919	Pit-N	125	478	61	132	11	4	2	27	37	26	.276	.332	.328	.660	94	0	-3	105	62	59	.659	31	15	-2
1920	Pit-N	137	550	78	154	19	15	4	32	45	28	.280	.341	.391	.732	109	7	6	105	48	74	.723	31	9	0
1921	Pit-N	147	632	100	204	23	17	3	42	41	19	.323	.364	.427	.791	105	8	5	103	48	94	.741	31	7	-1
1922	Pit-N	150	614	113	215	29	15	5	99	56	13	.350	.405	.471	.876	121	24	21	104	109	116	.894	21	2	
1923	Pit-N	123	499	79	149	18	7	1	54	43	15	.299	.355	.363	.718	93	-6	-4	97	93	65	.655	24		
1924	Pit-N	89	282	42	74	4	1	0	15	26	12	.262	.331	.284	.615	62	-12	-15	106	68	29	.577	10		
1925	Pit-N	66	126	15	30	7	0	0	8	7	8	.238	.278	.294	.572	45	-10	-11	102	77	10	.469	15		
1926	Pit-N	42	68	15	15	3	1	2	4	3	0	.221	.264	.382	.646	64	-3	-4	112	43	7	.604	2		
Total	11	1147	4192	629	1205	139	75	17	324	344	161	.287	.345	.369	.713	96	-2	-16	103	69	558	.685	182	68	

■ LYLE BIGBEE

Bigbee, Lyle Randolph "Al" b: 8/22/1893, Sweet Home, Ore. d: 8/5/42, Portland, Ore. BL/TR, 6', 180 lbs. Deb: 4/15/20

YEAR	TM/L	G	AB	R	H	2B	3B	HR	RBI	BB	SO	AVG	OBP	SLG	PRO	/A	BR	/A	PF	CHI	RC	TA	SB	CS	SBR
1920	Phi-A	38	75	5	14	2	0	1	8	9	12	.187	.282	.253	.536	45	-6	-5	94	128	6	.492	1	0	0
1921	Pit-N	5	2	0	0	0	0	0	0	0	1	.000	.000	.000	.000	-97	-1	-1	103	0	0	.000	0	0	0
Total	2	43	77	5	14	2	0	1	8	9	13	.182	.276	.247	.523	42	-7	-6	94	125	6	.476	1	0	0

■ ELLIOT BIGELOW

Bigelow, Elliot Allardice "Babe" or "Gilly" b: 10/13/1897, Tarpon Springs, Fla. d: 8/10/33, Tampa, Fla. BL/TL, 5'11", 185 lbs. Deb: 4/18/29

YEAR	TM/L	G	AB	R	H	2B	3B	HR	RBI	BB	SO	AVG	OBP	SLG	PRO	/A	BR	/A	PF	CHI	RC	TA	SB	CS	SBR
1929	Bos-A	100	211	23	60	16	0	1	26	23	18	.284	.357	.374	.732	87	-3	-4	102	108	28	.671	1	4	-2

■ CRAIG BIGGIO

Biggio, Craig Alan b: 12/14/65, Smithtown, N.Y. BR/TR, 5'11", 185 lbs. Deb: 6/26/88

YEAR	TM/L	G	AB	R	H	2B	3B	HR	RBI	BB	SO	AVG	OBP	SLG	PRO	/A	BR	/A	PF	CHI	RC	TA	SB	CS	SBR
1988	Hou-N	50	123	14	26	6	1	3	5	7	29	.211	.254	.350	.603	77	-5	-4	93	38	11	.566	6	1	1

FIGURE C.2: Craig Biggio's 1988 statistical record. Thorn and Palmer, *Total Baseball* (1989 ed.), p. 966

graded his abilities and makeup in an effort to provide a single number—60—that would represent his overall potential; the various calculations to determine a dollar value—$55,000–$60,000—for signing him, in part based upon an assessment of the habits of those who advised Biggio; and the specific language and phrases that Ferrick used to communicate what he saw to the front office, including judgments about his maturity and makeup.

There is, of course, much more to say about these documents and the infrastructure that created them. Taken together, they are a reminder of the ways both scorers and scouts have tried to make the evaluation of human beings—a process that is fallible, unpredictable, and messy—as reliable, useful, and powerful as possible. Baseball has long provided insight into American culture and history; it now can help us better understand the apparent triumph of data analytics by encouraging us to reassess how we know what we know about the game.

ACKNOWLEDGMENTS

When I began this project years ago, it was not clear what form—if any—the research would take. Nevertheless, I found a number of people in baseball willing to speak with me, point me in promising directions, and shape my understanding of scouting and scoring. In this regard, I am particularly indebted to Gary Gillette, Dave Smith, and John Thorn. Many others, however, took the time to speak with me as the project developed, often donating hours of time to help me understand this story. They include (in alphabetical order): Dick Cramer, John Dewan, (the late) Katy Feeney, Sean Forman, Tom Gilbert, Matt Gould, Howie Karpin, Kevin Kerrane, Sean Lahman, Phyllis Merhige, Pete Palmer, Laurel Prieb, Don Pries, Tom Ruane, Cory Schwartz, Jordan Sprechman, Stew Thornley, Tom Thress, Tom Tippett, Ted Turocy, (the late) Dave Vincent, and Craig Wright. I sincerely thank them all for their incredible generosity and apologize for any errors that stubbornly remain despite their best efforts. I also thank colleagues at Carnegie Mellon's History Department and New York University's Gallatin School of Individualized Study.

Once my "baseball project" started to talk and walk like a book, a number of colleagues and friends stepped in at a critical juncture to give me advice about just what sort of book it might become. In this capacity I happily thank Dan Bouk, Ann and Bruce Burkhardt, Will Deringer, Kent Murphy, Anna Evans Phillips, and Steven Shapin. I recognize it is a great imposition to ask someone to read one's own book in progress, and I am particularly thankful for their willingness to do so. I relied on an additional group of experts to help me navigate particular sections of the manuscript, including

Jeremy Blatter, Stephanie Dick, Gary Gillette, Dave Smith, Stew Thornley, and Craig Wright. Two anonymous readers also provided insightful and encouraging feedback. Al Bertrand recognized what I was trying to do with this book seemingly in the first minute of our conversation and he—alongside Mark Bellis, Sarah Vogelsong, Kristin Zodrow, and the rest of the team at Princeton University Press—has ably shepherded the book to completion.

A number of institutions and individuals helped with the initial research, including the staff at the Baseball Hall of Fame Library, the New York Public Library, the Harvard University Archives, the New-York Historical Society, the University of Delaware Archives, and the Wesleyan University Archives. I am particularly indebted to Rod Nelson, Jacob Pomrenke, and colleagues at the Society for American Baseball Research as well as to the pioneering interviews conducted by P. J. Dragseth, Kevin Kerrane, and members of the SABR Scouting and Official Scoring Committees; this book simply could not have existed without them. Permission to quote from archival materials and oral histories is gratefully acknowledged.

I am extraordinarily lucky to have a family that supports my research unfailingly, whether or not it involves baseball. Most importantly, Anna Evans Phillips has been a steady supporter of me during the time this book was under construction, even taking it upon herself to start keeping score during baseball games. She might have been humoring me initially, but her newfound love of scorekeeping reminds me just how wonderful it is to see the game with fresh eyes, rediscovering what matters. I may have first learned about baseball from my parents, but Anna has ensured that I never stop learning what baseball, and life, is really about.

ABBREVIATIONS USED IN NOTES

BHOF: Baseball Hall of Fame Library, Cooperstown, New York

UDARM: University of Delaware Archives and Records Management, Newark, Delaware

Mears Collection: Charles W. Mears Collection, Cleveland Public Library, Sports Research Center

HUA: Harvard University Archives, Cambridge, Massachusetts

NYHS: New-York Historical Society Manuscript Collection, New York

Spalding Collection: Spalding Baseball Collection, Manuscripts and Archives Division, New York Public Library, Astor, Lenox, and Tilden Foundations, New York

WUA: Special Collections and Archives, Olin Library, Wesleyan University, Middletown, Connecticut

NOTES

Introduction

1. Just how "modern" a revolution this is depends on one's perspective. Certainly books like Viktor Mayer-Schönberger and Kenneth Cukier, *Big Data: A Revolution that Will Transform How We Live, Work, and Think* (Boston: Houghton Mifflin, 2013) suggest it is a twenty-first-century phenomenon, while Jerry Muller's *The Tyranny of Metrics* (Princeton, NJ: Princeton University Press, 2018) locates the rise of "metric fixation" in the second half of the twentieth century. More historical-minded approaches note that Karl Pearson asked for "statistics on the table" over a century ago, Adolphe Quetelet believed the discovery of statistical regularities and patterns might promote social progress in the 1830s, and seventeenth-century philosophers wanted to align notions of reason with mathematical calculations. A good introduction to the history remains Gerd Gigerenzer, Zeno Swijtink, Theodore Porter, Lorraine Daston, John Beatty, and Lorenz Krüger, *The Empire of Chance: How Probability Changed Science and Everyday Life* (Cambridge: University of Cambridge Press, 1989).

2. Michael Lewis, *Moneyball: The Art of Winning an Unfair Game* (New York: Norton, 2003); and Steven Zaillian and Aaron Sorkin, *Moneyball*, dir. Bennett Miller (Columbia Pictures, 2011); Michael A. Roberto, "Billy Beane: Changing the Game," Case 9-305-120 (Boston: Harvard Business School Publishing, June 23, 2005); Jim Nussle and Peter Orszag, eds., *Moneyball for Government* (New York: Disruption Books, 2014); Dawinder S. Sidhu, "Moneyball Sentencing," *Boston College Law Review* 56, no. 2 (March 2015): pp. 671–731; Jim Soland, "Is 'Moneyball' the Next Big Thing in Education?: Educators Should Approach Early Warning Systems Thoughtfully and with Caution," *Phi Delta Kappan* 96, no. 4 (2014): p. 64; Anthony G. Vito and Gennaro F. Vito, "Lessons for Policing from *Moneyball*: The Views of Police Managers," *American Journal of Criminal Justice* 38, no. 2 (2013): p. 236.

3. For analytics and intuition, see Jay Liebowitz, ed., *Bursting the Big Data Bubble* (Boca Raton, FL: CRC Press, 2015), p. x; for objectivity and instinct, see Mayer-Schönberger and Cukier, *Big Data*, pp. 139–140; Lewis himself expresses these ideas in *Moneyball* as a "religious war" between enlightened rationality and superstition (pp. 17–18, 43, 287).

4. Daniel Rosenberg, "Data Before the Fact," in *"Raw Data" Is an Oxymoron*, ed. Lisa Gitelman, pp. 15–40 (Cambridge, MA: MIT Press, 2013); and Hans-Jörg Rheinberger, "Infra-Experimentality: From Traces to Data, From Data to Patterning Facts,"

History of Science 49 (2011): pp. 337–348, esp. p. 337. At least one commentator has noted that what we usually call "data" would be better termed "capta"—that which has been captured—since data entail an active process of selection rather than a passive state of being: Rob Kitchin, *The Data Revolution: Big Data, Open Data, Data Infrastructures and Their Consequences* (London: Sage, 2014), pp. 2–3.

5. John Thorn and Pete Palmer, *The Hidden Game of Baseball: A Revolutionary Approach to Baseball and Its Statistics* (Garden City, NY: Doubleday, 1984); for a more recent example repeating this division, see Keith Law, *Smart Baseball: The Story Behind the Old Stats that Are Ruining the Game, the New Ones that Are Running It, and the Right Way to Think about Baseball* (New York: William Morrow, 2017). It is worth noting that my story has very little overlap with the history of academic statistics, as usually understood. This is mainly because traditional concerns of statistics like modeling, sampling, and inference played relatively little role in the collection and analysis of baseball statistics until quite recently, but also because statistics as a discipline is increasingly divergent from data analysis. On this split, see the classic article by John W. Tukey, "The Future of Data Analysis," *Annals of Mathematical Statistics* 33, no. 1 (1962): pp. 1–67; as well as an update on the divide in Leo Breiman, "Statistical Modeling: The Two Cultures," *Statistical Science* 16, no. 3 (2001): pp. 199–231.

6. Some advocates of "big data" even assert its messiness and imprecision as a virtue—e.g., Mayer-Schönberger and Cukier, *Big Data*, pp. 32–49.

7. Paul N. Edwards, *A Vast Machine: Computer Models, Climate Data, and the Politics of Global Warming* (Cambridge: MIT Press, 2010), p. 22. Revealing the "infrastructure" is another way of expressing the need to "open the black box" to expose the inner workings of a machine, in this case a machine that generates baseball knowledge.

8. On these distinctions, I'm drawing from Steven Shapin, "The Sciences of Subjectivity," *Social Studies of Science* 42 (2012): pp. 170–184. An introduction to objectivity is Stephen Gaukroger, *Objectivity: A Very Short Introduction* (New York: Oxford, 2012). Historians have been interested in historicizing and analyzing objectivity since the early 1990s, producing a large literature on the topic, including Lorraine Daston and Peter Galison, *Objectivity* (New York: Zone Books, 2007); Theodore M. Porter, *Trust in Numbers: The Pursuit of Objectivity in Science and Public Life* (Princeton, NJ: Princeton University Press, 1995); Sheila Jasanoff, "The Practices of Objectivity in Regulatory Science," in *Social Knowledge in the Making*, ed. Charles Camic, Neil Gross, and Michèle Lamont, pp. 307–338 (Chicago: University of Chicago Press, 2011); Thomas F. Gieryn, "Objectivity for These Times," *Perspectives on Science* 2, no. 3 (1994): pp. 324–349; Theodore M. Porter, "Objectivity as Standardization: The Rhetoric of Impersonality in Measurement, Statistics, and Cost-Benefit Analysis," *Annals of Scholarship* 9, nos. 1–2 (1992): pp. 19–60; Lorraine Daston, "Objectivity and the Escape from Perspective," *Social Studies of Science* 22, no. 4 (1992): pp. 597–618; and Peter Dear, "From Truth to Disinterestedness in the Seventeenth Century," *Social Studies of Science* 22, no. 4 (1992): pp. 619–631.

9. In effect, I mean to follow both scorers and scouts as others have followed scientists—looking at their actual practices, tools, and institutions rather than only at the way they talk about or represent their work publicly. For the classic account

in science studies, see Bruno Latour, *Science in Action: How to Follow Scientists and Engineers Through Society* (Cambridge: Harvard University Press, 1987); for the ways focusing on practice might inform our understanding of objectivity, see Andy Pickering, "Objectivity and the Mangle of Practice," *Annals of Scholarship* 8, nos. 3–4 (1991): pp. 409–426.

10. Cathy O'Neil, *Weapons of Math Destruction: How Big Data Increases Inequality and Threatens Democracy* (New York: Crown, 2016), esp. p. 17.

11. This feature—and its meaning—has been described by Hans-Jörg Rheinberger in far more detail with scientific instrumentation: see Rheinberger, "Infra-Experimentality"; idem, "Epistemic Objects/Technical Objects," in *Epistemic Objects*, Max Planck Institute for the History of Science Preprint no. 374, ed. Uljana Feest, Hans-Jörg Rheinberger, and Günter Abel, pp. 93–98 (Berlin: Max Planck, 2009); and idem, *Toward a History of Epistemic Things: Synthesizing Proteins in the Test Tube* (Palo Alto, CA: Stanford University Press, 1997). For a broader look at this approach to the history of technology, see David Edgerton, *The Shock of the Old: Technology and Global History since 1900* (Oxford: Oxford University Press, 2007); Lisa Gitelman, *Paper Knowledge: Toward a Media History of Documents* (Durham, NC: Duke University Press, 2014); and Henry Petroski, *The Pencil: A History of Design and Circumstance* (New York: Knopf, 1992 [1989]).

12. Though there is more than adequate evidence that clubs must develop both scouting and analytic capabilities, the World Series victory of Biggio's longtime club, the Houston Astros, in 2017 was tellingly covered in newspaper articles that featured headlines about the moment when "analytics conquered" Major League Baseball but in their particulars were more focused on the success of the club's amateur scouts. See, e.g., Dave Sheinin, "Astros' World Series Win May Be Remembered as the Moment Analytics Conquered MLB for Good," *Washington Post* (November 2, 2017), https://www.washingtonpost.com/sports/astros-world-series-win-may-be-remembered-as-the-moment-analytics-conquered-mlb-for-good/2017/11/02/ac62abaa-bfec-11e7-97d9-bdab5a0ab381_story.html (accessed November 22, 2017).

13. The topic of foreign players is a parallel story, as amateur scouts increasingly are responsible for searching for talent worldwide, especially given the relatively low cost of finding many foreign players. Foreign scouting is beyond the scope of this work, though many of the skills and practices honed for discovering domestic players are also used internationally; the main differences are the institutions by which foreign players' labor is controlled, often from a very young age, including "camps" and "academies," especially in Latin America. See Daniel A. Gilbert, *Expanding the Strike Zone: Baseball in the Age of Free Agency* (Amherst: University of Massachusetts Press, 2013), pp. 107–135; Rob Ruck, "Baseball's Recruitment Abuses," *Americas Quarterly* (Summer 2011), http://americasquarterly.org/node/2745 (accessed April 18, 2018); and idem, *Raceball: How the Major Leagues Colonized the Black and Latin Game* (Boston: Beacon, 2011).

14. As of July 2016, these reports were located at http://scouts.baseballhall.org, but they were taken offline in December 2016 and are slated as of 2018 to be eventually relaunched as part of the Hall's PASTIME initiative at http://collection.baseballhall.org.

15. Republished as Bill James, "The Epic of Craig Biggio," *Slate* (February 25, 2008), http://www.slate.com/articles/sports/sports_nut/2008/02/the_epic_of_craig_biggio.html (accessed December 21, 2016). Two decades earlier, James himself had developed a metric to assess Hall of Fame worthiness that combined qualitative and quantitative features: Bill James, *The Politics of Glory: How Baseball's Hall of Fame Really Works* (New York: Macmillan, 1994). For a more detailed, if occasionally hagiographic, account of Biggio's career (including the process by which he was scouted and signed), see David Siroty, *The Hit Men and The Kid Who Batted Ninth: Biggio, Valentin, Vaughn and Robinson Together Again in the Big Leagues* (Lanham, MD: Diamond Communications, 2002).

Chapter 1: The Bases of Data

1. Major League Baseball (MLB) lists Cobb at .385 and Lajoie at .384 (http://mlb.mlb.com/stats/league_leaders.jsp#season=1910 [accessed January 13, 2017]), whereas Baseball Reference lists Cobb at .383 (http://www.baseball-reference.com/leaders/batting_avg_top_ten.shtml [accessed January 13, 2017]). For the controversy, see Rick Huhn, *The Chalmers Race: Ty Cobb, Napoleon Lajoie, and the Controversial 1910 Batting Title that Became a National Obsession* (Lincoln: University of Nebraska Press, 2014).

2. Baseball Reference, "Core Purpose Statement," May 1, 2004, http://www.baseball-reference.com/about/sources.shtml (accessed September 19, 2016).

3. Interview with Sean Forman, December 10, 2014.

4. Interview with Sean Lahman, April 10, 2015. Technically, the database was originally posted in 1996 on a different website, http://vivanet.com/~lahmans/baseball.html, but by 1998 baseball1.com was the only place to find it.

5. *Feist Publications, Inc. v. Rural Tel. Service Co.*, 499 U.S. 340 (1991), http://laws.findlaw.com/us/499/340.html (accessed April 9, 2015).

6. Electronic computers and Cold War military science were nearly inextricable: Paul N. Edwards, *The Closed World: Computers and the Politics of Discourse in Cold War America* (Cambridge: MIT Press, 2006).

7. For Palmer's account of these sources, see John Thorn and Pete Palmer, eds., *Total Baseball* (New York: Warner Books, 1989), pp. 681–686.

8. Alan Schwarz, *The Numbers Game: Baseball's Lifelong Fascination with Statistics* (New York: Thomas Dunne, 2004), pp. 92–109.

9. *Spalding's Base Ball Guide and Official League Book for 1890* (Chicago: Spalding, 1890), p. 183 (also online at https://archive.org/details/spaldingsbasebal1890chic [accessed August 14, 2017]).

10. Thorn and Palmer, *Total Baseball*, pp. 679–694; see also Pete Palmer, "The History of Total Baseball and Pete Palmer's Baseball Databases," Society for American Baseball Research, February 12, 2005, http://sabr.org/cmsFiles/PalmerDatabaseHistory.pdf (accessed January 13, 2017).

11. Lars Heide, *Punched-Card Systems and the Early Information Explosion, 1880–1945* (Baltimore: Johns Hopkins University Press, 2009).

12. David Alan Grier, *When Computers Were Human* (Princeton, NJ: Princeton University Press, 2005); for gendered computing, see Janet Abbate, *Recoding Gender:*

Women's Changing Participation in Computing (Cambridge: MIT Press, 2012); and
Jennifer S. Light, "When Computers Were Women," *Technology and Culture* 40, no.
3 (1999): pp. 455–483.

13. Interview with Pete Palmer, December 19, 2014.

14. "Database," *Oxford English Dictionary*, 3d ed. (Oxford: Oxford University
Press, 2012); see also Thomas Haigh, "How Data Got Its Base: Information Storage Software in the 1950s and 1960s," *IEEE Annals of the History of Computing* 31,
no. 4 (2009): pp. 6–25. On the significance of early databases providing access to
knowledge, see Geoffrey C. Bowker, *Memory Practices in the Sciences* (Cambridge:
MIT Press, 2005), though unlike Paul Otlet's early-twentieth-century Mundaneum,
Palmer did not make any attempt to create a "universal" database: Alex Wright,
Cataloging the World: Paul Otlet and the Birth of the Information Age (Oxford: Oxford University Press, 2014); on midcentury card-based projects storing human data,
see Rebecca Lemov, *Database of Dreams: The Lost Quest to Catalog Humanity* (New
Haven, CT: Yale University Press, 2015); on the history of card catalogs as machines,
see Markus Krajewski, *Paper Machines: About Cards and Catalogs, 1548–1929* (Cambridge: MIT Press, 2011 [2002]).

15. Computer History Museum, Exhibition on Supercomputers, http://www
.computerhistory.org/revolution/supercomputers/10/33 (accessed August 14, 2017).

16. Interview with Pete Palmer, December 19, 2014.

17. See Davids's letters of March 3, 1971, April 16, 1971, and June 28, 1971, at
"SABR's 40th Anniversary," Society for American Baseball Research, http://sabr.org
/40years (accessed June 9, 2016).

18. See SABR's own account at http://sabr.org/about/history (accessed June 9,
2016).

19. In this regard, the history of baseball research is not all that different from the
history of science itself, which was for many centuries not usually something one was
paid to do directly, but more akin to a hobby or side interest. Even as science itself
became a job or vocation, it was often practiced using resources created for other
purposes. For an account of this tension in twentieth-century America, see Steven
Shapin, *The Scientific Life: A Moral History of a Late Modern Vocation* (Chicago: University of Chicago Press, 2008).

20. Interview with Palmer, December 19, 2014.

21. Thorn and Palmer, *Total Baseball*, pp. 679–694; Palmer, "History of Total
Baseball."

22. More recently, the *Feist* copyright decision was used in a decision followed
closely by those who care deeply about the use of baseball statistics. In the 2006 case
C.B.C. Distribution and Marketing v. Major League Baseball Advanced Media, the
issue was (in part) whether independent groups might use the names and statistics of
players in their own fantasy games without licensing them from the Players' Association. Again, the courts confirmed that players' names and records are not subject to
copyright. See *C.B.C. Distribution and Marketing v. Major League Baseball Advanced
Media*, 443 F.Supp. 2d 1077 (2006).

23. "Data friction" is a term taken from Paul N. Edwards, *A Vast Machine: Computer Models, Climate Data, and the Politics of Global Warming* (Cambridge: MIT

Press, 2010); the notion of making data, particularly quantifiable data, both stable and mobile is expressed in the now-classic notion of an "immutable mobile": Bruno Latour, *Science in Action: How to Follow Scientists and Engineers Through Society* (Cambridge: Harvard University Press, 1987), esp. pp. 227–228.

24. Technically, the licensing of Palmer's database was through Gary Gillette's companies 24-7 Baseball and Hidden Game Sports.

25. Interview with Forman, December 10, 2014.

26. The distinctions are even blurrier in this case because when Palmer began his work, there was no such thing as an "expert" in baseball statistics, and despite the proliferation of baseball analytics since, few consider the collection of baseball statistics analogous to more formal scientific endeavors. For a historical analysis of data-driven science, see Bruno J. Strasser, "Data-driven Sciences: From Wonder Cabinets to Electronic Databases," *Studies in History and Philosophy of Biological and Biomedical Sciences* 43 (2012): pp. 85–87; on the blurring of expert and vernacular knowledge in science outside of the usual places, see Christine von Oertzen, Maria Rentetzi, and Elizabeth S. Watkins, "Finding Science in Surprising Places: Gender and the Geography of Scientific Knowledge, Introduction to 'Beyond the Academy: Histories of Gender and Knowledge,'" *Centaurus* 55 (2013): pp. 73–80; and for blurring in the usual places, see Jeffrey R. Young, "Crowd Science Reaches New Heights," *Chronicle of Higher Education* (May 28, 2010), http://www.chronicle.com/article/The-Rise-of-Crowd-Science/65707 (accessed July 17, 2017). For examples of the involvement of lay persons in scientific data collection, see Jeremy Vetter, "Introduction: Lay Production in the History of Scientific Observation," *Science in Context* 24, no. 2 (2011): pp. 127–141; Christine von Oertzen, "Science in the Cradle: Milicent Shinn and Her Home-Based Network of Baby Observers, 1890–1910," *Centaurus* 55 (2013): pp. 175–195; Emmanuel Didier, "Counting on Relief: Industrializing the Statistical Interviewer During the New Deal," *Science in Context* 24, no. 2 (2011): pp. 281–310; Lorraine Daston, "The Immortal Archive: Nineteenth-Century Science Imagines the Future," in *Science in the Archives: Pasts, Presents, Futures*, ed. Lorraine Daston, pp. 159–182 (Chicago: University of Chicago Press, 2017); and Etienne S. Benson, "A Centrifuge of Calculation: Managing Data and Enthusiasm in Early Twentieth-Century Bird Banding," *Osiris* 32 (2017): pp. 286–306.

27. Andrew Warwick, "The Laboratory of Theory or What's Exact about the Exact Sciences?" in *The Values of Precision*, ed. M. Norton Wise, pp. 311–351 (Princeton, NJ: Princeton University Press, 1995).

Chapter 2: Henry Chadwick and Scoring Technology

1. Lorraine Daston, "Hard Facts," in *Making Things Public: Atmospheres of Democracy*, ed. Bruno Latour and Peter Weibel, pp. 680–683 (Cambridge: MIT Press, 2005). More broadly, see Lorraine Daston and Katharine Park, *Wonders and the Order of Nature, 1150–1750* (New York: Zone Books, 1998); and Mary Poovey, *A History of the Modern Fact: Problems of Knowledge in the Sciences of Wealth and Society* (Chicago: University of Chicago Press, 1988). Tellingly, though numbers came to be closely associated with general observations and facts, there was nothing that

required this connection, and indeed eighteenth-century general observations included both numbered and unnumbered forms: J. Andrew Mendelsohn, "The World on a Page: Making a General Observation in the Eighteenth Century," in *Histories of Scientific Observation*, ed. Lorraine Daston and Elizabeth Lunbeck, pp. 397–420 (Chicago: University of Chicago Press, 2011).

2. Allen Guttmann, *From Ritual to Record: The Nature of Modern Sports* (New York: Columbia University Press, 1978), esp. pp. 47–55, 95–116.

3. John Thorn and Pete Palmer, eds., *Total Baseball* (New York: Warner Books, 1989).

4. Andrew J. Schiff, *"The Father of Baseball": A Biography of Henry Chadwick* (Jefferson, NC: McFarland, 2008).

5. Christopher Hamlin, *Public Health and Social Justice in the Age of Chadwick, Britain, 1800–1854* (Cambridge: Cambridge University Press, 1998), p. 85.

6. Edwin's letter quoted in Frederick Ivor-Campbell, "Vita: Henry Chadwick," *Harvard Magazine* 90 (September–October 1987): pp. 60–61 (quotation on p. 60); and cited in Jules Tygiel, *Past Time: Baseball as History* (New York: Oxford University Press, 2000), p. 19. Henry's letter of November 13, 1888, to Edwin Chadwick, appears as no. 463 in Chadwick Papers, University College London, Special Collections, London, United Kingdom.

7. Chadwick Scorebooks, 1858–1907, Henry Chadwick Papers, Spalding Collection; quotation from Henry Chadwick, *Lawn Tennis Score Book* (1882), unnumbered first page, Spalding Collection, reprinted in idem, *The Lawn Tennis Manual for 1885* (New York: Spalding, 1885), pp. 52–53.

8. "Every club": Henry Chadwick, ed., *1861 Beadle's Dime Base-ball Player* (New York: Beadle, 1861), p. 54; Diagrams: Chadwick's Scorebooks, Spalding Collection; also in Henry Chadwick, ed., *De Witt's Baseball Guide for 1869* (New York: De Witt, 1869), p. 10. For Chadwick's description of the scorer, see Henry Chadwick, *The Game of Baseball: How to Learn It, How to Play It, and How to Teach It* (New York: George Munro, 1868), pp. 20–21, 58–59.

9. See Adam Chadwick, *A Portrait of Lord's: 200 Years of Cricket History* (London: Scala Arts & Heritage, 2013), p.118; Samuel Britcher, *A Complete List of All the Grand Matches of Cricket* (London: T. Craft, 1793).

10. Henry Chadwick, ed., *Base Ball Player's Book of Reference* (New York: J. C. Haney, 1867), p. vi.

11. Knickerbocker Baseball Club Game Books, Spalding Collection; one source for tracking the changing use of the statistic "hands lost" or "outs" is the scrapbook kept by William Rankin, who worked for the *New York Clipper*: Mears Baseball Scrapbook, vol. 1, Mears Collection.

12. Game Scores and Club Administration, 1858–1859, Records of the Lawrence Base Ball Club, HUD 9500, Harvard University Archives, Cambridge, Massachusetts; Knickerbocker Baseball Club Game Books, Spalding Collection; Alert Baseball Club Scorebook, 1859–1860, NYHS.

13. Agallian Base Ball Club Records, 1865–1869, Collection 1000-43, Box 1, Folders 4–5, WUA; Atlantic Baseball Club: Game Book, October 21, 1855–October 29, 1868, Spalding Collection.

14. See John Thorn, "The Baseball Press Emerges," *Our Game* (blog), n.d., http: //ourgame.mlblogs.com/2011/10/20/the-baseball-press-emerges (accessed June 15, 2016); and R. Terry Furst, *Early Professional Baseball and the Sporting Press: Shaping the Image of the Game* (Jefferson, NC: McFarland, 2014).

15. "Reports of their play": Mears Baseball Scrapbook, vol. 4, p. 1, Mears Collection; "Ball Play," *New York Clipper* (December 10, 1859), p. 268.

16. "Ball Play," *New York Clipper* (December 10, 1859), p. 268.

17. "Ball Play: A Correct Score of a Base Ball Match," *New York Clipper* (January 14, 1860), p. 308.

18. For a representative example, see "Averages of the Atlantic Club for 1861," *New York Clipper* (November 16, 1861), p. 247. Chadwick's scorebooks are preserved in the Spalding Collection. They also detail how his scoring system did not emerge fully developed but was honed over the years, beginning with the recording only of outs and runs and moving to records of "additional statistics" and "full" scores.

19. Chadwick, *Game of Baseball*, pp. 45, 67–68. Emphasis in original.

20. See, e.g., Chadwick, *De Witt's Base-Ball Guide for 1869*, p. 9.

21. "Ball Play: Batting Averages," *New York Clipper* (March 17, 1860), p. 380. The article announcing the 1868 Clipper Medals (from early 1869) can be found in Mears Baseball Scrapbook, vol. 4, p. 57, Mears Collection.

22. Henry Chadwick, ed., *De Witt's Baseball Guide for 1872* (New York: De Witt, 1872), pp. 54–97; quotation from Henry Chadwick, ed., *De Witt's Baseball Guide for 1873* (New York: De Witt, 1873), pp. 70–71.

23. Chadwick, *1861 Beadle's Baseball Guide*, pp. 58–60; "generally adopted": Chadwick, *Game of Baseball*, pp. 62–63.

24. "First improvement": Chadwick, *Game of Baseball*, p. 11; Chadwick, *1861 Beadle's Baseball Guide*, pp. 56–57.

25. Chadwick, *Base Ball Player's Book of Reference* (1867), pp. 67–69.

26. Chadwick, *Game of Baseball*, p. 66.

27. Chadwick Scrapbook, vol. 1, p. 3 [1860?], Spalding Collection.

28. Chadwick, *De Witt's Base-Ball Guide for 1869*, p. 82; cf. Chadwick, *Game of Baseball*, p. 62.

29. "Chances": pasted into Chadwick's Scorebooks, vol. 5, Spalding Collection; left vs. right fielder: Henry Chadwick, ed., *1860 Beadle's Dime Base-ball Player* (New York: Beadle, 1860), pp. 25–26.

30. The December 1868 rules are bound in vol. 3 of Chadwick's Scorebooks, Spalding Collection. The initial discussion of this play appeared in the 1870s—see, e.g., Henry Chadwick, *Base Ball Manual* (London: Routledge, 1874), p. 80—though by the end of the next decade, Chadwick had revised the statistic to 3 in 5 rather than 26 in 30: Henry Chadwick, *How to Play Base Ball* (Chicago: Spalding, 1889), p. 134.

31. For home runs vs. singles, see "The Atlantics Averages for 1868," Chadwick Scrapbooks, vol. 1, Spalding Collection; "equally detailed analysis": Chadwick, *Game of Baseball*, pp. 11–12.

32. Chadwick, *De Witt's Base-Ball Guide for 1869*, pp. 83–85.

33. For rational accounting as a rhetorical strategy, see Bruce G. Carruthers and Wendy Nelson Espeland, "Accounting for Rationality: Double-Entry Bookkeeping

and the Rhetoric of Economic Rationality," *American Journal of Sociology* 97, no. 1 (1991): pp. 31–69.

34. Lisa Gitelman, ed., *"Raw Data" Is an Oxymoron* (Cambridge: MIT Press, 2013), plate 2; Scott A. Sandage, *Born Losers: A History of Failure in America* (Cambridge: Harvard University Press, 2005).

35. Quotation from Chadwick, *Game of Baseball*, p. 69. Quetelet's vision (along with that of his contemporaries) for social improvement through statistical record-keeping is traced in Gerd Gigerenzer, Zeno Swijtink, Theodore Porter, Lorraine Daston, John Beatty, and Lorenz Krüger, *The Empire of Chance: How Probability Changed Science and Everyday Life* (Cambridge: Cambridge University Press, 1989), pp. 37–69. On Quetelet's scientific work, see Kevin Donnelly, *Adolphe Quetelet, Social Physics and the Average Men of Science, 1796–1874* (London: Pickering and Chatto, 2015).

36. Chadwick, *Game of Baseball*, pp. 38–43.

37. Chadwick, *Base Ball Player's Book of Reference* (1867), pp. 69–70.

38. In historian Dan Bouk's typology, his statistics were simultaneously used as "data doubles"—standing in for a particular player by labeling him with a particular number—and as "data aggregates"—lumping together disparate players in order to show stability or change over time across teams or leagues: Dan Bouk, "The History and Political Economy of Personal Data over the Last Two Centuries in Three Acts," *Osiris* 32 (2017): pp. 85–108.

39. Jules Tygiel, *Past Time: Baseball as History* (New York: Oxford University Press, 2000), p. 25.

40. Henry Chadwick, ed., *The Base Ball Player's Book of Reference* (New York: J. C. Haney, 1866), pp. vii–viii, 61–63; also "Baseball Season of 1866," *Sunday Mercury* (December 30, 1866).

41. John Thorn, *Baseball in the Garden of Eden: The Secret History of the Early Game* (New York: Simon & Schuster, 2011), esp. p. 87.

42. E.g., "efficient scorer": Chadwick, *De Witt's Base-Ball Guide for 1869*, p. 78; Cincinnati Red Stockings Scorebooks, 1868–1870, vol. 1, games of September 28, September 29, and October 7, 1868, Spalding Collection; M. J. Kelly: Chadwick, *De Witt's Base-Ball Guide for 1869*, p. 9; Harry Wright's Pocket Base Ball Score Book, folder "H.U.B.B.C. Scorebook of the Harvard Nine, 1876–1877," Records of Organized Baseball at Harvard, 1876–1945, Box 2, HUD 9500, Harvard University Archives, Cambridge, Massachusetts. See also the Harry Wright Papers in the Spalding Collection.

43. Mutual's 1871 scores are kept in vol. 4 of Chadwick's Scorebooks, Spalding Collection; May 28, 1881, game found in vol. 7 of Chadwick's Scorebooks, Spalding Collection. Scorebooks function similarly to standards more generally: there is always a range of behavior from "near compliance" to "outright defiance," and processes of standardization, like those of scorebook creation, should always be considered incomplete and iterative, subject to future revision in order to increase rates of compliance: Susan Leigh Star and Martha Lampland, "Reckoning With Standards," in *Standards and their Stories: How Quantifying, Classifying, and Formalizing Practices Shape Everyday Life*, pp. 3–24 (Ithaca, NY: Cornell University Press, 2009), esp. pp. 14–15.

44. Chadwick, *De Witt's Base-Ball Guide for 1869*, pp. 77–78.

45. See, e.g., Correspondence Scrapbooks, vol. 1, Knickerbocker Base Ball Club Files, Spalding Collection.

46. Chadwick, *De Witt's Base-Ball Guide for 1869*, p. 82.

47. Ian Hacking, *The Taming of Chance* (Cambridge: Cambridge University Press, 1990), pp. 2–3. The metaphor of avalanches, or deluges, of data has been criticized by Bruno Strasser, who notes such images mask the extensive labor involved: "The data deluge, far from being a natural phenomenon or the result of a technological revolution, was achieved only slowly and in the teeth of resistance by those who envisioned how publicly available big data could transform the production of scientific knowledge." Bruno J. Strasser, "The 'Data Deluge': Turning Private Data into Public Archives," in *Science in the Archives: Pasts, Presents, Futures*, ed. Lorraine Daston, pp. 185–202 (Chicago: University of Chicago Press, 2017), esp. p. 187.

Chapter 3: Official Scoring

1. Bill Plaschke, "Hit or Error?" *Los Angeles Times* (July 11, 1993), pp. C3, C10.

2. In 2017, Vice President of Special Projects Laurel Prieb took over the management of scorers, though he acknowledged that the system was in good enough shape that he was planning to build on Merhige's work rather than start afresh; interview with Laurel Prieb, July 11, 2017.

3. Interview with Phyllis Merhige, January 30, 2015; John McMurray, interview with Ben Trittipoe, October 10, 2016, SABR Official Scoring Committee files, https://sabr.org/research/official-scoring-research-committee (accessed April 25, 2018).

4. Interview with Merhige, January 30, 2015.

5. See, e.g., "Convention Rules for 1859," printed in Alert Baseball Club Scorebook, 1859–1860, NYHS; "gentlemanly conduct": Henry Chadwick, ed., *1860 Beadle's Dime Base-ball Player* (New York: Beadle, 1860), p. 30; "pretty severe trial": "Base Ball: Instructions in Scoring," *New York Clipper* (March 23, 1861), p. 388; "fond of statistical work": Henry Chadwick, *The Game of Base Ball: How to Learn It, How to Play It, and How to Teach It* (New York: George Munro, 1868), p. 162.

6. Henry Chadwick, *The Base Ball Player's Book of Reference* (New York: J. C. Haney, 1866), pp. 30, 59–60. On gentility and reliability, see Steven Shapin, *A Social History of Truth: Civility and Science in Seventeenth-Century England* (Chicago: University of Chicago Press, 1994), esp. pp. 65–126.

7. "Easy proposition": C. P. Stack, "Pleasure and Profit of Keeping Score," *Baseball Magazine* 13 (May 1914): pp. 61–64 (quotation on p. 61); "no easy matter": George L. Moreland, "Need of a Better Scoring System," *Baseball Magazine* 2 (November 1908): pp. 15–16 (quotation on p. 15); Henry Chadwick, *Base Ball Player's Book of Reference* (New York: J. C. Haney, 1867), p. 124.

8. Harvard vs. Cincinnati scorecard, June 12, 1869, in folder "Baseball 1869," as well as scorecards in folders "Baseball 1870," "Baseball 1873," "Baseball 1874," and "Baseball 1876," HUD9000, Harvard University Baseball Collection, box 1, Harvard University Archives, Cambridge, Massachusetts.

9. See Alex J. Haas, "A Woman Official Scorer," *Baseball Research Journal* 5 (1976), http://research.sabr.org/journals/woman-official-scorer (accessed April 25, 2018); "Magnates Seek Chas. Williams for Secretary," *Chicago Daily Tribune* (January 13, 1918), p. A3; The Pilgrim, "Woman as Official Scorer," *Washington Post* (May 14, 1905), p. E10. One notable exception was Susan Fornoff of Oakland, who was also the first woman to be named to the board of directors of the Baseball Writers' Association of America; more recently, Marie-Claude Pelland-Marcotte served as Toronto's official scorer. See Stew Thornley's profile of Susan Fornoff for SABR at https://sabr.org/node/41612 (accessed August 15, 2017).

10. National League of Professional Base Ball Clubs, *Constitution and Playing Rules of the National League of Professional Base Ball Clubs* (Philadelphia: Reach and Johnston, 1876), pp. 20–21; "proper man": "Games and Pastimes: Further Facts Concerning the New Base-Ball League," *Chicago Daily Tribune* (February 13, 1876), p. 12.

11. National League of Professional Base Ball Clubs, *Constitution and Playing Rules of the National League of Professional Base Ball Clubs* (Chicago: Spalding, 1877), pp. 38–40. Forms were originally mentioned not in the constitution or rules, but in the *Chicago Daily Tribune*'s "Games and Pastimes."

12. See, e.g., *Spalding's Base Ball Guide* (Chicago: Spalding, 1880), p. 36.

13. See comments of then–National League secretary John Heydler, quoted in M.V.B. Lyons, "The Problem of Official Scoring," *Baseball Magazine* 10 (November 1912): pp. 55–60 (quotation on pp. 59–60); as well as Moreland, "Need of a Better Scoring System," pp. 15–16.

14. F. C. Lane, "Behind the Scenes in Organized Baseball," *Baseball Magazine* 10 (January 1913): pp. 49–54.

15. See, e.g., Weber's essay on bureaucratic authority, reprinted in Max Weber, *Sociological Writings*, ed. Wolf Heydebrand, pp. 59–107 (New York: Continuum, 1994).

16. *Spalding's Official Baseball Guide* (Chicago: Spalding, 1887), pp. 136–137.

17. Ernest J. Lanigan, "Origin of Statistics," in *Baseball Register*, pp. 33–36 (St. Louis: Sporting News, 1942).

18. Henry P. Edwards, "History of Base Ball Writers Association of America," in *Spalding's Official Baseball Guide*, ed. John Foster, pp. 43–44 (New York: American Sports Publishing Company, 1939) (quotation on p. 43).

19. "Gain control": "Baseball Writers Organize," *New York Times* (October 15, 1908), p. 7; "more uniform": "Big Leagues in Session," *Boston Daily Globe* (December 10, 1908), p. 5.

20. "The Shifting Records," *Baseball Magazine* 17 (August 1916): pp. 24–26 (quotation on p. 26).

21. The 1921 transition is noted in Bill Shannon, *Official Scoring in the Big Leagues* (New York: Sports Museum Press, 2006), p. 4; I have failed to find decisive evidence to support the claim. Clearly it is the case that BBWAA scorers were used exclusively by the end of the decade and had been for a while; see, e.g., Rud Rennie, "Major Leagues Do Little in Joint Meetings," *New York Herald Tribune* (December 13, 1929), p. 33; and Edwards, "History of Base Ball Writers," p. 44.

22. Moreland, "Need of a Better Scoring System," pp. 15–16.

23. M.V.B. Lyons, "The Problem of Official Scoring," *Baseball Magazine* 10 (November 1912): pp. 55–60.

24. J. Ed Grillo, "Sporting Comment," *Washington Post* (January 1, 1910), p. 8; Frank Vaccaro, "Origin of the Modern Pitching Win," *Baseball Research Journal* (Spring 2013), http: //sabr.org/research/origin-modern-pitching-win (accessed July 28, 2016).

25. The information in this paragraph is from Rick Huhn, *The Chalmers Race: Ty Cobb, Napoleon Lajoie, and the Controversial 1910 Batting Title that Became a National Obsession* (Lincoln: University of Nebraska Press, 2014).

26. Moreland, "Need of a Better Scoring System," pp. 15–16. Nearly every major newspaper covered the controversy, but see esp. "Lynch Suggests Changes in Scorers," *New York Times* (October 21, 1910), p. 8; "Taylor the Storm Center of Big Baseball Sessions," *Washington Post* (December 11, 1910), p. S1; Joe S. Jackson, "Johnson Is the Only National Taylor Wants for Boston," *Washington Post* (December 13, 1910), p. 8; and idem, "M'Aleer as Magnate," *Washington Post* (December 15, 1910), p. 8; see also Huhn, *Chalmers Race*, p. 155.

27. "Club Scorers are Faint Hearted Lot," *Hartford Courant* (September 5, 1915), p. 18.

28. Irving Vaughan, "Ban Silences Talk of War on Landis," *Chicago Daily Tribune* (December 13, 1922), p. 26. This episode has entered scoring lore. See Shannon, who calls it perhaps the "most important game" in the twentieth century for official scoring as a profession, in *Official Scoring*, p. 5; Dick Young, "They Call That a Hit in Detroit," in *This Great Game*, ed. Doris Townsend, pp. 246–255 (New York: Rutledge, 1971), pp. 252–254; and Dave Anderson, "Rose vs. Cobb: Hit or Error?" *New York Times* (September 10, 1985), p. B10, among others.

29. "Gotham Scribes Do Not Favor Cobb's 400 Average," *Washington Post* (December 10, 1922), p. 62.

30. Vaughan, "Ban Silences Talk of War," p. 26; Fred Lieb, *Baseball As I Have Known It* (New York: Coward, McCann, and Geoghegan, 1977), p. 70. Lieb describes the Ty Cobb phantom hit episode as taking place in August rather than in May, as suggested by the bulk of the evidence.

31. See Hugh Fullerton, "Plan Changes in League Scoring of Ball Games," *Chicago Daily Tribune* (April 14, 1923), p. 18, "League Presidents May Name Official Scorers," *Washington Post* (December 27, 1922), p. 12; and for the quotation, "Johnson Tells Scorers to Put House in Order," *Boston Daily Globe* (December 23, 1922), p. 7. Owners' control over players through their control of official scorers is memorably recounted in Lieb, *Baseball As I Have Known It*, esp. p. 67.

32. The change was so minor in actual consequence that one reporter concluded after the announcement that nothing "startling or even interesting happened": Rennie, "Major Leagues Do Little," p. 33. Other reporters called it an "experiment," though it was continued: "Thompson Is Named Nats' Official Scorer," *Washington Post* (April 14, 1930), p. 13.

33. Dan Daniel, "Frick Opens Fire on Slipshod Scoring," *Sporting News* (August 3, 1949), pp. 1, 6.

34. J. G. Taylor Spink, ed., *Baseball Guide and Record Book: 1950* (St. Louis: Sporting News, 1950), p. 558. Copies of subsequent rules can be found in the yearly guidebooks for this period.

35. See, e.g., Steve Jacobson, "Sportswriter as Scorer: An Error All the Way," *Newsday* (June 17, 1979), p. E4; Harold Kaese, "Ban Writers as Scorers," *Boston Globe* (August 4, 1963), p. 65.

36. Young quoted in "Sports in Mid-Winter," *Washington Post* (February 17, 1889), p. 2; Lynch quoted in Joe S. Jackson, "Johnson Is Only National Taylor Wants for Boston," *Washington Post* (December 13, 1910); Victor Jones, "Scorer (Once 'Mr. X') Has Formidable Job," *Boston Daily Globe* (May 23, 1948), p. C1; "In the Dugout with Rumill: Official Scoring," *Christian Science Monitor* (February 27, 1963), p. 7.

37. Allan Lewis, "Full-Time Scorers?: Majors Facing Problem," *Sporting News* (November 6, 1976), p. 28; see also Leonard Koppett, "A 4-Step Plan to Improving Official Scoring," *Sporting News* (August 29, 1981), p. 8.

38. Jack O'Connell, "Hall of Fame Writer Jack Lang Dead at 85," *MLB.com* (January 25, 2007), http: //m.mlb.com/news/article/1786612/ (accessed May 26, 2015); "smallest particle of say": "The Four Toughest Calls: Is There a Conflict in the Press Box?" *Los Angeles Times* (July 6, 1978), p. F10; Kuhn quoted in Jack Craig, "The Official Scorers," *Boston Globe* (August 13, 1978), p. 92; Arthur Friedman, with Joel H. Cohen, *The World of Sports Statistics: How the Fans and Professionals Record, Compile and Use Information* (New York: Atheneum, 1978), esp. p. 120.

39. Jack Lang, "Writers Can Keep Jobs as Scorers," *Sporting News* (December 22, 1979), p. 44; Phil Elderkin, "Baseball's Scribes Field Conflict-of-Interest Charges," *Christian Science Monitor* (August 7, 1980), p. 16.

40. "Few Writers Are Permitted to Score Now," *Los Angeles Times* (August 8, 1982), p. D11.

41. Letter to editor of the *New York Clipper* from a "Boston reader," November 21, 1873, in Chadwick Scrapbooks, vol. 1, Henry Chadwick Papers, Spalding Collection. The original article, "Base-Hits and Earned Runs," was pasted into vol. 5 of Chadwick's Scorebooks, after the 1873 season scores, Spalding Collection, and published as Henry Chadwick, *Base Ball Manual* (London: Routledge and Sons, 1874), pp. 78–83.

42. "Instructions to Scorers," from November[?] 1876, Mears Baseball Scrapbook, vol. 4, p. 203, Mears Collection; National League of Professional Base Ball Clubs, *Constitution and Playing Rules of the National League of Professional Base Ball Clubs* (Chicago: Spalding and Bros., 1877), pp. 38–39.

43. T. H. Murnane, *How to Play Baseball* (1904–1908 editions), MVFB p.v.45, nos. 1–5, Spalding Collection; J. M. Cummings, *How to Score: A Practical Textbook for Scorers of Baseball Games, Amateur and Expert* (New York: American Sports Publishing Company, 1911), pp. 3, 6–10.

44. Cummings, *How to Score*, pp. 6–7, 15–25.

45. I. E. Sanborn, "How to Score a Baseball Game," *Chicago Daily Tribune* (March 31, 1912), p. C2.

46. "The National League: Instructions to Official Scorers 1920," MVFB n.c. 161, pp. 1–4, Spalding Collection.

47. H. G. Fisher, *How to Score* (Chicago: Fisher, 1925); *The Scorecard Newsletter*, GV879.S36, BHOF Newsletters, no. 1 (Apr. 1969) to no. 17 (July 16, 1970), BHOF (Gammon was likely the official scorer for the nearby Chattanooga Lookouts).

48. J. G. Taylor Spink, ed., *Baseball Guide and Record Book: 1943* (St. Louis: Sporting News, 1943), p. 57; Young, "They Call That a Hit," p. 248.

49. Murnane, *How to Play Baseball* (1904), p. 109.

50. These examples are taken from *Official Baseball Rules: 2017 Edition* (New York: Office of the Commissioner of Baseball, 2017), on pp. 75, 99, 107, 120, 127, 145.

51. Jon Leonoudakis, interview with Ed Munson, June 14, 2016, SABR Official Scoring Committee files, https: //sabr.app.box.com/s/0sxvdoux6pyb595jb8k9id6qlj wswz5u/file/73337481165 (accessed July 20, 2017).

52. "Baseball: How They Rank," *Chicago Tribune* (November 8, 1981), p. C10; see also Jerome Holtzman, "A Baseball Bonanza: Secret Ratings of All the Players," *Chicago Tribune* (November 8, 1981), p. C1.

53. Aaron Lucchetti, "Scoring Baseball Games Is Thankless Task at Best," *Wall Street Journal* (September 24, 2003), http: //www.wsj.com/articles/SB106435740 377622600 (accessed April 25, 2018), citing Merhige.

54. The precise process within MLB's offices is not specified, though the basics of the appeal to the executive vice president for baseball operations are outlined in Rule 19(d) of the 2018 edition of Major League Baseball, *Official Professional Baseball Rules Book*, p. 96, https: //registration.mlbpa.org/pdf/MajorLeagueRules.pdf (accessed April 25, 2018). Numbers taken from Lucchetti, "Scoring Baseball Games Is Thankless"; Bill Christine, "An Official Scorer's Call Isn't Always the Last Word," *New York Times* (July 7, 2012), p. SP11; personal interview with Merhige, January 30, 2015; and Wayne Strumpfer's interview with Chuck Dybdal, July 2, 2016, SABR Official Scoring Committee Files, https: //sabr.app.box.com/s/0sxvdoux6pyb595jb8k9id6qljwswz5u /file/73173227565 (accessed July 20, 2017).

55. Interview with Stew Thornley, June 11, 2015; Leonoudakis, interview with Munson, June 14, 2016.

56. Though the league avoids specifics, one scorer noted that in 2015, there were 410 appeals of scoring decisions, only 107 of which made it to the executive vice president, and only 56 of which were actually changed: Strumpfer, interview with Dybdal, July 2, 2016; interview with Thornley, June 11, 2015.

57. These examples come from Strumpfer, interview with Dybdal, July 2, 2016; and McMurray, interview with Trittipoe, October 10, 2016, respectively; "subjective": interview with Thornley, June 11, 2015.

58. "De facto" is Thornley's term: interview with Thornley, June 11, 2015; "like a family": Bill Nowlin's interview with Mike Shalin, April 28, 2016, SABR Official Scoring Committee Files, https: //sabr.app.box.com/s/0sxvdoux6pyb595jb8k9id6qljwswz5u /file/72993238857 (accessed April 25, 2018); for an example of a past overruling of consensus opinion, see Leonoudakis, interview with Munson, June 14, 2016.

59. See introduction, note 8, for an introduction to the scholarly literature on objectivity.

60. "Editorials," *Baseball Magazine* 11 (June 1913): pp. 13–14 (quotation on p. 13).

Chapter 4: From Project Scoresheet to Big Data

1. Keith Law, *Smart Baseball: The Story Behind the Old Stats that Are Ruining the Game, the New Ones that Are Running It, and the Right Way to Think about Baseball* (New York: William Morrow, 2017), esp. pp. 242–245. For a general introduction to the conception of big data as novel, see Rob Kitchin, "Big Data, New Epistemologies and Paradigm Shifts," *Big Data & Society* 1, no. 1 (April–June 2014): pp. 1–12; and idem, *The Data Revolution: Big Data, Open Data, Data Infrastructures and Their Consequences* (London: Sage, 2014); for a typical example of the boosterism that surrounds the "big data revolution," see Viktor Mayer-Schönberger and Kenneth Cukier, *Big Data: A Revolution that Will Transform How We Live, Work, and Think* (Boston: Houghton Mifflin, 2013). Historicizing big data tends to have the effect of deflating some of its claims of novelty—recognizing, for example, that "big science" was itself a transformation that predated the rapid increase in data storage capacities of the late twentieth century, and that people created "large" collections of facts, such as wonder cabinets and museums, for the purpose of new insights centuries ago; among many, see Elena Aronova, Christine von Oertzen, and David Sepkoski, "Introduction: Historicizing Big Data," *Osiris* 32 (2017): pp. 1–17; and Bruno J. Strasser, "Collecting Nature: Practices, Styles, and Narratives," *Osiris* 27 (2012): pp. 303–340.

2. Bill James, *The Bill James Baseball Abstract 1984* (New York: Ballantine Books, 1984), pp. 250–252. The account of Project Scoresheet and STATS in this chapter relies upon Schwarz's *Numbers Game* as well as interviews with Dick Cramer (December 17, 2014), John Dewan (May 6, 12, and 14, 2015), Gary Gillette (October 2, 2014 and August 18, 2017), and Dave Smith (October 31, 2014).

3. See, e.g., G. R. Lindsey, "Statistical Data Useful for the Operation of a Baseball Team," *Operations Research* 7, no. 2 (1959): pp. 197–207. For George Lindsey, as well as Harlan and Eldon Mills, Earnshaw Cook, and others, see Alan Schwarz, *The Numbers Game: Baseball's Lifelong Fascination with Statistics* (New York: St. Martin's, 2004), esp. pp. 67–91.

4. See 1984's American League and National League accounts in Project Scoresheet Account Books, BHOF Serials.

5. Efforts like Wikipedia suggest that the struggle of volunteer initiatives isn't a matter of people's general unwillingness to volunteer their time to enable others to make money; rather, the relevant issue is perhaps the tension between volunteering in a larger data—even scientific—project for the good of all (even if a few profit indirectly) and simply working without pay for a profitable entity. For more on these tensions and their relationship to data collection efforts, see the literature on crowdsourced data gathering in chapter 1, note 25.

6. Bill James, *The Bill James Historical Baseball Abstract* (New York: Villard Books, 1985); Seymour Siwoff, Steve Hirdt, and Peter Hirdt, *The 1985 Elias Baseball Abstract* (New York: Macmillan, 1985); John Thorn and Pete Palmer, *The Hidden Game of Baseball: A Revolutionary Approach to Baseball and Its Statistics* (Garden City, NY: Doubleday, 1984); and John Thorn and Pete Palmer, eds., *Total Baseball* (New York: Warner, 1989).

7. STATS, "Computerized Play by Play Scoring for Baseball," April 1, 1990, p. 16, private collection of Dick Cramer.

8. Interview with Smith, October 31, 2014; also quoted in Stefan Fatsis, "Batty?: A Man Tries to Log Every Play Made in Baseball—David Smith Feels Past Is Key to National Pastime; Maris Loses an RBI," *Wall Street Journal* (April 24, 2002), p. A1.

9. Interview with David Vincent, February 13, 2015. As with Palmer's work in chapter 1, this precisely demonstrates the idea of "data friction": Paul N. Edwards, *A Vast Machine: Computer Models, Climate Data, and the Politics of Global Warming* (Cambridge: MIT Press, 2010).

10. Bruno J. Strasser, "Data-Driven Sciences: From Wonder Cabinets to Electronic Databases," *Studies in History and Philosophy of Biological and Biomedical Sciences* 43 (2012): pp. 85–87.

11. Interview with Tom Ruane, February 6, 2015.

12. Interview with Tom Thress, December 5, 2014.

13. On Sherri Nichols's involvement, see Ben Lindbergh, "The Sabermetric Movement's Forgotten Foremother," *Ringer* (February 20, 2018), https://www.theringer.com/mlb/2018/2/20/17030428/sherri-nichols-baseball-sabermetric-movement (accessed February 26, 2018).

14. See, e.g., David Kalist and Stephen J. Spurr, "Baseball Errors," *Journal of Quantitative Analysis in Sport* 2, no. 4 (2006): pp. 1–20 (quotation on p. 6).

15. Interview with Tom Tippett, April 26, 2015.

16. "Total Sports Celebrates First Anniversary by Relaunching www.TotalBaseball.com," *PR Newswire*, March 31, 1998, p. 1. KOZ Sports was an online pioneer, run by the same group that had put the *Raleigh* (North Carolina) *News and Observer* online as nando.net; the merger with Total Sports' online business presented an opportunity to start putting real-time data online across many different sporting events.

17. "Company News: Quokka Adding Total Sports for about $130 Million," *New York Times* (July 22, 2000), p. C3.

18. Betsy Schiffman, "Disaster of the Day: Quokka Sports," *Forbes* (February 13, 2011), http://www.forbes.com/2001/02/13/0213disaster.html (accessed September 27, 2017). Quokka's rapid and complete failure was well documented at the time; see, e.g., Janelle Brown, "Auction of the Damned," *Salon.com* (July 27, 2001), https://www.salon.com/2001/07/27/auction/ (accessed November 22, 2017).

19. Interview with John Thorn, May 13, 2015. Of course, this insight about the physicality of the book was not limited to this particular one—as noted in chapter 1, the Macmillan *Baseball Encyclopedia* from 1969 was not just a huge book priced at today's equivalent of $175, but also the first published book to be typeset electronically. For a retrospective account, see John Thorn's interview with David Neft at the SABR meeting on July 1, 2017, http://sabr.org/convention/sabr47-committees, and https://sabr.app.box.com/s/xhzxdjx60qp8me0b755o7lfwoy2xrd5g (accessed August 2, 2017).

20. Interview with Gillette, August 18, 2017.

21. "Quokka Sports and Major League Baseball Advanced Media to Partner Through Multi-Year Technology License and Live Scoring Agreement," *PR Newswire* (February 23, 2001).

22. Interview with Cory Schwartz and Matthew Gould, May 8, 2015.

23. Andrew Blum, *Tubes: A Journey to the Center of the Internet* (New York: Ecco, 2012), p. 164.

24. In 2017, Disney acquired majority ownership of BAMTech, a spinoff company focused on video streaming, though MLBAM continued to maintain analytical operations and data collection for MLB: "The Walt Disney Company to Acquire Majority Ownership of BAMTech," BusinessWire (August 8, 2017), https: //www .businesswire.com/news/home/20170808006428/en/ (accessed April 25, 2018).

25. Interview with Schwartz and Gould, May 8, 2015.

26. Bruce Schoenfeld, "Can New Technology Bring Baseball's Data Revolution to Fielding?" *New York Times* (September 30, 2016), http: //www.nytimes.com/2016 /10/02/magazine/can-new-technology-bring-baseballs-data-revolution-to-fielding .html (accessed January 4, 2017).

27. There is a growing literature on the possible difference being "born digital" after about 1970 makes: Geoffrey C. Bowker and Susan Leigh Star, *Sorting Things Out: Classification and Its Consequences* (Cambridge: MIT Press, 1999), p. 7; Bruno J. Strasser and Paul N. Edwards, "Big Data Is the Answer . . . But What Is the Question?," *Osiris* 32 (2017): pp. 328–345; W. Patrick McCray, "The Biggest Data of All: Making and Sharing a Digital Universe," *Osiris* 32 (2017): pp. 243–263; and Lisa Gitelman, *Paper Knowledge: Toward a Media History of Documents* (Durham, NC: Duke University Press, 2014), esp. pp. 111–135.

28. Law, *Smart Baseball*, p. 259; Hallam Stevens, "A Feeling for the Algorithm: Working Knowledge and Big Data in Biology," *Osiris* 32 (2017): pp. 151–174. For the argument that computers are often built upon existing practices, see Jon Agar, "What Difference Did Computers Make?," *Social Studies of Science* 36, no. 6 (2006): pp. 869–907; for an alternative perspective on the transformative power of modern computing, see Stephanie Aleen Dick, "After Math: (Re)configuring Minds, Proof, and Computing in the Postwar United States" (Ph.D. diss., Harvard University, 2014). Perhaps it is a matter of timing—Stevens, in "Feeling for the Algorithm," finds the claim of continuity more compelling for electronic computing pre-1970 than in later years.

29. This argument about the interconnections of instrument, observation, and theory was made persuasively in the context of twentieth-century physics: Peter Galison, *Image and Logic: A Material Culture of Microphysics* (Chicago: University of Chicago Press, 1997), pp. 781–803.

Chapter 5: The Practice of Pricing the Body

1. More detail on the connection between Seton Hall's team and Major League Baseball is in David Siroty, *The Hit Men and The Kid Who Batted Ninth: Biggio, Valentin, Vaughn and Robinson Together Again in the Big Leagues* (Lanham, MD: Diamond Communications, 2002). See also Seton Hall's tributes to Biggio's college career at http: //www.shupirates.com/sports/m-basebl/spec-rel/010615aaa.html (accessed April 14, 2016) and http: //shupirates.com/news/2015/9/30/Craig_Biggio_Seton _Hall_s_First_Baseball_Hall_of_Famer.aspx?path=baseball (accessed April 23, 2018).

2. Allan Simpson, "Top 10 Largest Signing Bonuses, 1987 Draft," *Perfect Game* (March 30, 2007), http: //www.perfectgame.org/Articles/View.aspx?article=206 (accessed January 19, 2017).

3. Gib Bodet quoted in P. J. Dragseth, *Major League Scout: Twelve Thousand Baseball Games and Six Million Miles* (Jefferson, NC: McFarland, 2014), p. 51; Alexander and Wagner quoted in Kevin Kerrane, *Dollar Sign on the Muscle: The World of Baseball Scouting* (New York: Beaufort Books, 1984), pp. 41, 229, respectively.

4. Lee Lowenfish, "29 Years and Counting: A Visit with Longtime Cubs Scout Billy Blitzer," *Baseball Research Journal* (Spring 2011), https: //sabr.org/research/29 -years-and-counting-visit-longtime-cubs-scout-billy-blitzer (accessed January 4, 2017); see also Andy Martino, "For the Cubs' Ailing Longtime Scout Billy Blitzer, It Might Be Now or Never to Win a World Series Ring," *New York Daily News* (October 1, 2016), http: //www.nydailynews.com/sports/baseball/ailing-baseball-scout -chicago-cubs-win-year-article-1.2810895 (accessed January 4, 2017).

5. Kerrane, *Dollar Sign on the Muscle*; P. J. Dragseth, ed., *Eye for Talent: Interviews with Veteran Baseball Scouts* (Jefferson, NC: McFarland, 2010); Jim Sandoval and Bill Nowlin, eds., *Can He Play?: A Look at Baseball Scouts and Their Profession* (Phoenix: Society for American Baseball Research, 2011).

6. John J. Evers and Hugh S. Fullerton, *Touching Second: The Science of Baseball* (Chicago: Reilly and Britton, 1910), pp. 33–34. For an early example of "scout" in use, see Joe S. Jackson, "Detroit Wants Wood and Carrigan of Boston," *Washington Post* (December 14, 1910), p. 8; Kerrane, *Dollar Sign on the Muscle*, provides some of this history as well.

7. See Dan Levitt, "Development of the Yankees Scouting Staff," Jim Sandoval, "Larry Sutton," and Brian McKenna, "Cy Slapnicka," in Sandoval and Nowlin, *Can He Play?*, pp. 4–6, 19–22, 26–29; also Dragseth, *Eye for Talent*, p. 6.

8. Kerrane, *Dollar Sign on the Muscle*, pp. 5–9.

9. Alexander quoted in Dragseth, *Eye For Talent*, p. 22; Levitt, "Development of the Yankees Scouting Staff." Also see Dan Levitt, *Ed Barrow: The Bulldog Who Built the Yankees' First Dynasty* (Lincoln: University of Nebraska Press, 2008).

10. The signing of Feller is told, among many places, in Kerrane, *Dollar Sign on the Muscle*, p. 12. Lee Lowenfish, *The Imperfect Diamond: A History of Baseball's Labor Wars*, rev. ed. (Lincoln: University of Nebraska Press, 2010 [1980]), notes that this was not surprising, even if it was extreme—and Slapnicka avoided punishment only because Feller and his family were happy with the arrangements (p. 120).

11. Lowenfish, *Imperfect Diamond*, pp. 106–107.

12. Ron Smiley and Jim Sandoval, "Pop Kelchner," and Jim Sandoval, "Charley Barrett: The King of Weeds," in Sandoval and Nowlin, *Can He Play?*, pp. 8–12 (86 signings had been attributed to Kelchner and 66 to Barrett as of 2011).

13. Bruce Chadwick, *Baseball's Hometown Teams: The Story of the Minor Leagues* (New York: Abbeville Press, 1994), p. 102; Kerrane, *Dollar Sign on the Muscle*, pp. 9–11; Neil J. Sullivan, *Minors: The Struggles and the Triumph of Baseball's Poor Relation from 1876 to the Present* (New York: St. Martin's, 1990).

14. Lowenfish, *Imperfect Diamond*, pp. 115–126. The rules of the draft were tweaked repeatedly over time, as was the compensation for these picks.

15. Chadwick, *Baseball's Hometown Teams*, pp. 130–136.

16. Cincinnati Baseball Clubs, "Scout Contracts" and "Correspondence" folders, BA MSS 51, boxes 4–5, BHOF; other figures are from the annual *Spalding's Official Baseball Guide*; Kerrane, *Dollar Sign on the Muscle*, pp. 8 and 19 is consistent with these estimates.

17. McKenna, "Cy Slapnicka."

18. Gumpert: "General Scouting Bulletin for 1958," Randy Gumpert Scouting Collection, BA MSS 232, box 2, "1958 Scouting Manual—NYY" folder, BHOF; Gardner: Dragseth, *Eye for Talent*, p. 45.

19. For a firsthand account, see Dave Baldwin, "A Player's View of Scouts in the 1950s," and Bill Nowlin, "Johnny Pesky, On Signing with the Red Sox," both in Sandoval and Nowlin, *Can He Play?*, pp. 82–86; Dragseth, *Eye for Talent*, pp. 111 (Thompson) and 178 (McAlister); Yankees: "General Scouting Bulletin for 1958"; Hugh Alexander and Daniel Austin, "Baseball's Road Scholar: From the Dirt Roads to the Interstates [2009]," MFF 430, "Baseball's Road Scholar" folder, BHOF.

20. "Harry O'Donnell Scouting Note Books BL-3948.97," BA MSS 74, BHOF. Even as a general manager who could make personnel decisions directly, Branch Rickey kept written records of all players he had seen, reports now located in the Library of Congress digital collections: Branch Rickey Papers, 1890–1969, MSS 37820, Manuscript Division, Library of Congress, Washington, DC, http: //hdl.loc.gov/loc .mss/eadmss.ms998023 (accessed April 17, 2018).

21. "Pittsburgh Pirates Operations Manual," MFF 525, BHOF.

22. Ibid., p. 6.

23. Ibid., pp. 5–6.

24. Lee Lowenfish, "'Don't Get Too Domestic Out There': The Baseball Life and Insights of Herb Stein (1917–2010)," in Sandoval and Nowlin, *Can He Play?*, pp. 143–146; and Stan Hart, *Scouting Reports: The Original Reviews of Some of Baseball's Greatest Stars* (New York: Macmillan, 1995), pp. 31–33.

25. Rod Carew, report no. 1464 (May 1964), Scouting Report Collection, BHOF.

26. Reprinted in Hart, *Scouting Reports*, p. 34.

27. Mike McCormick, report no. 1538 (November 1955), Scouting Report Collection, BHOF. Other descriptions are culled from a number of different reports.

28. *Eye for Talent* is the title of Dragseth's book. For the relevant literature on cowboys, see, e.g., William Cronin, *Nature's Metropolis: Chicago and the Great West* (New York: Norton, 1991), esp. pp. 219–220; and for truckers, Shane Hamilton, *Trucking Country: The Road to America's Wal-Mart Economy* (Princeton, NJ: Princeton University Press, 2008).

29. "Pittsburgh Pirates Operations Manual"; "Texas Rangers Scouting Manual (Revised January 1990)," Joseph S. Branzell Collection, BA MSS 114, p. 4, BHOF.

30. George Gmelch and J. J. Weiner, *In the Ballpark: The Working Lives of Baseball People* (Washington, DC: Smithsonian Institute Press, 1998), pp. 36–37.

31. Mark Winegardner, *Prophet of the Sandlots: Journeys with a Major League Scout* (New York: Atlantic Monthly Press, 1990), 9–10, 57; Joe E. Palmer, *Old Baseball Scout and His Players: Horsehide and Hollywood* (Beverly Hills, CA: Remlap Publishing, 1987); David V. Hanneman, *Diamonds in the Rough: The Legend and Legacy*

of Tony Lucadello (Austin, TX: Diamond Books, 1989); Hugh Alexander and David Austin, "Baseball's Road Scholar," MFF 430, BHOF. For a clear example of the typical structure, see Jim Russo, *Super Scout: Thirty-Five Years of Major League Scouting* (Chicago: Bonus Books, 1992).

32. "Ancient Mariner" and "Old Scout": Bill Plaschke, "Scout's Honor," in *The Best American Sports Writing 2010*, ed. Peter Gammons, pp. 304–207 (Boston: Houghton Mifflin Harcourt, 2010); "good baseball men": Dragseth, *Eye for Talent*, p. 97, quoting Dick Wilson; Collins: Kerrane, *Dollar Sign on the Muscle*, p. 1.

33. Dragseth, *Eye for Talent*, pp. 216 (Espy), 183 (McAlister), 57–58 (McIntyre), 206 (Rizzo).

34. Ibid., p. 164 (Mock); Kerrane, *Dollar Sign on the Muscle*, p. 116 (McLaughlin).

35. Branzell Collection, Box 1, BHOF.

36. Jim Sandoval, "Bessie and Roy Largent: Baseball's Only Husband and Wife Scouting Team," and Shawn Selby, "Edith Houghton," both in Sandoval and Nowlin, *Can He Play?*, pp. 22–24, 63–66; Paul Vitello, "Edith Houghton, Rare Woman among Baseball Scouts, Dies at 100," *New York Times* (February 16, 2003), p. D8; Art Stewart crediting his wife Donna as "another scout": Art Stewart, with Sam Mellinger, *The Art of Scouting: Seven Decades Chasing Hopes and Dreams in Major League Baseball* (Olathe, KS: Ascend, 2014), p. 22. For more recent examples of female scouts, see Noah Frank, "Before 'Pitch': A Woman Breaks Through Another of Baseball's Glass Ceilings," *WTOP News* (September 16, 2016), http: //wtop.com/mlb/2016/09/pitch -woman-breaks-another-baseballs-glass-ceilings/ (accessed September 30, 2016); and Tim Booth, "Hopkins 1st Full-time Female Baseball Scout in Over 50 Years," Associated Press (August 18, 2017), https: //apnews.com/52b38ee2da9347678e952405959ba18e (accessed September 8, 2017).

37. Rory Costello, "Sam Hairston," in Sandoval and Nowlin, *Can He Play?*, pp. 127–132; and Buck O'Neil, with Steve Wulf and David Conrads, *I Was Right on Time* (New York: Simon & Schuster, 1996), p. 203; Clary quoted in Dragseth, *Eye for Talent*, pp. 68–69. For a recent summary of the role of race in scouting, see Bill Brink, "MLB Has a Diversity Problem, and It Goes Beyond the Rosters," *Pittsburgh Post-Gazette* (August 21, 2017), http: //www.post-gazette.com/sports/pirates/2017/08/21/mlb -minority-managers-dave-roberts-dusty-baker-rick-renteria/stories/201708190011 (accessed September 8, 2017).

38. Alan Schwarz, "Baseball America's Scouting Dictionary," *Baseball America* (November 14, 2006), http: //www.baseballamerica.com/online/majors/column /2006/262830.html (accessed October 12, 2015); Kerrane, *Dollar Sign on the Muscle*, pp. 87–101 also gives a scouting glossary; Lucadello quoted in Winegardner, *Prophet of the Sandlots*, p. 38.

39. "Tall": Hartzell, report no. 1251 (September 9, 1974); "big": Knapp, report no. 1120 (April 18, 1975); "long": Cambria, report no. 2043 (March 9, 1969); "compact": Dewitt, report no. 5485 (April 2, 2004); "good athletic": Hurdle, report no. 5156 (March 31, 1975); "has the carriage": Kennedy, report no. 5018 (May 11, 1968); "good agile": Picciolo, report no. 1563 (January 1, 1975); "live body": Long, report no. 48 (April 4, 1994); "some softness": Meadows, report no. 92 (April 20, 1994);

"strong; rugged": North, report no. 2597 (June 4, 1968), all in Scouting Report Collection, BHOF.

40. "Washington Senators Scouts Manual (1961)," Branzell Collection, Box 4, p. 9, BHOF; Kerrane, *Dollar Sign on the Muscle*, pp. 58 ("little righties") and 235 (Nickels); Dragseth, *Eye for Talent*, p. 164 (Mock).

41. "Carries himself": Picciolo, report no. 1563 (January 1, 1975); "similar to Woodward": Heintzelman, report no. 2473 (April 21, 1968); "similar to Grich": Krenchicki, report no. 5428 (April 6, 1975); "something of a Todd Helton": Dewitt, report no. 5485 (April 2, 2004); "reminds of Lofton": Long, report no. 48 (April 4, 1994), all in Scouting Report Collection, BHOF.

42. Bill Shanks, *Scouts' Honor: The Bravest Way to Build a Winning Team* (New York: Sterling and Ross, 2005), p. 330.

43. Dragseth, *Eye for Talent*, p. 63 (Clary); Allan Simpson, ed., *The Baseball Draft: The First 25 Years, 1965–1989* (Durham, NC: American Sports Publishing, 1990), pp. 19, 197.

44. Bodet, *Major League Scout*, p. 163 (in a chapter titled "Off to the Races"); a similar metaphor was used to describe the importance of intangibles by Ralph DiLullo, quoted in Paul Post, "Foresight 20/20: The Life of Baseball Scout Ralph DiLullo" (1995), p. 76, GV865.D47.P67, BHOF.

45. Guy Griffith and Michael Oakeshott, *A New Guide to the Derby* (London: Faber and Faber, 1947 [originally published in 1936 as *A Guide to the Classics*]), esp. pp. 7, 49.

46. For Atlanta: Shanks, *Scouts' Honor*, p. 25; and Gmelch and Weiner, *In the Ballpark*, p. 31; "bloodlines": Schwarz, "Baseball America's Scouting Dictionary."

47. For examples of "physical description": Hartzell, report no. 1251 (September 9, 1974); "glasses": Krenchicki, report no. 5428 (April 6, 1975); Knapp, report no. 1120 (April 18, 1975); and Picciolo, report no. 1563 (January 1, 1975), all in Scouting Report Collection, BHOF; Lucadello: Winegardner, *Prophet of the Sandlots*, pp. 104, 241–242.

48. "Physical defects": Hartzell, report no. 1251 (September 9, 1974), Scouting Report Collection, BHOF; "no known injuries" and "physical defects": Hart, *Scouting Reports*, p. 34; "strong durable type": Krenchicki, report no. 5428 (April 6, 1975); "elbow chips": Pastore, report no. 1560 (May 20, 1975), both in Scouting Report Collection, BHOF.

49. For report with "married" category, see (among many) Amalfitino, report no. 1645 (April 1953), Scouting Report Collection, BHOF; e.g., "too old": "Washington Senators Scouts Manual (1961)," p. 9.

50. Alexander and Austin, "Baseball's Road Scholar," p. 49; Kerrane, *Dollar Sign on the Muscle*, p. 15.

51. For this use of "nationality," see Gibson, report no. 192 (July 15, 1956); Robinson, report no. 199 (June 8, 1953), both in Scouting Report Collection, BHOF.

52. For a representative Philadelphia report, see Boone, report no. 997 (June 18, 1969), Scouting Report Collection, BHOF; Hamilton quoted in Kerrane, *Dollar Sign on the Muscle*, pp. 108–109.

53. Winegardner, *Prophet of the Sandlots*, p. 35.

54. Bill Nowlin, "Lou Gorman: 'You Don't Win Without Good Scouts': A GM Looks at Scouting," in Sandoval and Nowlin, *Can He Play?*, pp. 116–127; Dragseth, *Eye for Talent*, pp. 177 (McAlister) and 215 (Espy); Lennie Merullo, interview with Bob Bencks, November 10, 1994, SABR Oral History Files, https: //oralhistory.sabr .org/interviews/merullo-lennie-1994/ (accessed April 26, 2018).

55. O'Neil, with Wulf and Conrads, *I Was Right on Time*, pp. 202–203; Doyle: Neal Mackertich, "Scouting Jack Doyle," in Sandoval and Nowlin, *Can He Play?*, pp. 6–8; Lucadello: Winegardner, *Prophet of the Sandlots*, pp. 203–204; Davis: Kerrane, *Dollar Sign on the Muscle*, p. 140.

56. Lucadello: Hanneman, *Diamonds in the Rough*, p. 104; cf. Winegardner, *Prophet of the Sandlots*, p. 42; Mock: Dragseth, *Eye for Talent*, p. 164.

57. "Chicago Cubs Scouting Manual," BA MSS 81, Dan Austin Collection, Box 1, pp. 13–14, BHOF; "Philadelphia Phillies Scouting Manual [1978]," BA MSS 101, Tony Lucadello Collection, folder 17, p. 2-1, BHOF.

58. "Texas Rangers Scouting Manual [1990]," pp. 14–15.

59. Ibid., p. 24: Lucadello: Hanneman, *Diamonds in the Rough*, p. 103; and Winegardner, *Prophet of the Sandlots*, pp. 97, 238–239.

60. See Gib Bodet, "National Cross Checking," in Sandoval and Nowlin, *Can He Play?*, pp. 150–156, who credits Al Campanis for the expression "normal development of a player"; Cronin: Gmelch and Weiner, *In the Ballpark*, p. 31; "Philadelphia Phillies Scouting Manual [1978]," p. 2-1.

61. Dragseth, *Eye for Talent*, p. 60.

62. "Philadelphia Phillies Scouting Manual [1978]," p. 11-3.

63. J. M. Ward, "Is the Ball-Player a Chattel?," *Lippincott's Monthly Magazine* 40 (August 1887): pp. 310–319; on his influence, see Robert F. Burk, *Never Just a Game: Players, Owners, and American Baseball to 1920* (Chapel Hill: University of North Carolina Press, 1994), pp. 94–115.

64. Letter quoted in Leonard Koppett, "Kuhn Denies Flood's Request to 'Free' Him for Other Offers Besides Phils'," *New York Times* (December 31, 1969), p. 32; cited in Daniel A. Gilbert, *Expanding the Strike Zone: Baseball in the Age of Free Agency* (Amherst: University of Massachusetts Press, 2013), p. 42.

65. Bell, report no. 187 (August 27, 1955), Scouting Report Collection, BHOF; Walter Johnson, *Soul by Soul: Life Inside the Antebellum Slave Market* (Cambridge: Harvard University Press, 1999), esp. p. 17.

66. O'Neil, with Wulf and Conrads, *I Was Right on Time*, p. 203; Al Goldis with John Wolff, *How to Make Pro Baseball Scouts Notice You: An Insider's Guide to Big League Scouting* (New York: Skyhorse, 2009), pp. 63–75. On self-tracking, see Kate Crawford, Jessa Lingel, and Tero Karppi, "Our Metrics, Ourselves: A Hundred Years of Self-Tracking from the Weight Scale to the Wrist Wearable Device," *European Journal of Cultural Studies*, 18, nos. 4–5 (2015): pp. 479–496; and Deborah Lupton, *The Quantified Self* (Cambridge: Polity, 2016); on the history of the related practice of self-archiving, see Rebecca Lemov, "Archives-of-Self: The Vicissitudes of Time and Self in a Technologically Determinist Future," in *Science in the Archives: Pasts,*

Presents, Futures, ed. Lorraine Daston, pp. 247–270 (Chicago: University of Chicago Press, 2017).

67. Frank McGowan, "Judgment—The First Ingredient in Scouting," box 1, folder 9, BA MSS 114, p. 3, Joseph S. Branzell Collection, BHOF.

68. See, e.g., Kansas City: Mason, report no. 5065 (April 20, 1968); Houston: Dauer, report no. 1360 (May 13, 1974); Milwaukee: Woodard, report no. 122 (April 23, 1994); Chicago White Sox: Cruz, report no. 269 (1994); "money expected": Norwood, report no. 1555 (May 15, 1972), all in Scouting Report Collection, BHOF.

69. Biggio, report no. 31 (March 17, 1987); Biggio, report no. 1236 (March 15, 1987); Biggio, report no. 223 (March 18, 1987), all in Scouting Report Collection, BHOF; Simpson, "Top 10 Largest Signing Bonuses"; see also baseball-reference.com for more detailed lists.

70. Lucien Karpik, *Valuing the Unique: The Economics of Singularities* (Princeton, NJ: Princeton University Press, 2010). Of course, governments *are* particularly concerned with applying economic logic to the value of human life in other contexts. For a recent review of that literature, see Katherine Hood, "The Science of Value: Economic Expertise and the Valuation of Human Life in U.S. Federal Regulatory Agencies," *Social Studies of Science* 47, no. 4 (2017): pp. 441–465.

71. For the interconnections of observation and reading/writing, see Lorraine Daston, "Taking Note(s)," *Isis* 95 (2004): pp. 443–448; for the historical role of notetaking and "paper tools" in the development of classification schemes and scientific data, see, among others, Boris Jardine, "State of the Field: Paper Tools," *Studies in History and Philosophy of Science Part A* 64 (August 2017): pp. 53–63; Michael Bennett, "Note-Taking and Data-Sharing: Edward Jenner and the Global Vaccination Network," *Intellectual History Review* 20, no. 3 (2010): pp. 415–432; Volker Hess and J. Andrew Mendelsohn, "Sauvages' Paperwork: How Disease Classification Arose from Scholarly Note-Taking," *Early Science and Medicine* 19 (2014): pp. 471–503; idem, "Case and Series: Medical Knowledge and Paper Technology, 1600–1900," *History of Science* 48 (2010): pp. 287–314; Staffan Müller-Wille and Isabelle Charmantier, "Lists as Research Technologies" *Isis* 103, no. 4 (2012): pp. 743–752; and Richard Yeo, *Notebooks, English Virtuosi, and Early Modern Science* (Chicago: University of Chicago Press, 2014).

72. On the gap between the image and reality of a scientific life, see Steven Shapin, *The Scientific Life: A Moral History of a Late Modern Vocation* (Chicago: University of Chicago Press, 2008).

Chapter 6: Measuring Head and Heart

1. The scene is in Michael Lewis, *Moneyball: The Art of Winning an Unfair Game* (New York: Norton, 2004 [2003]), p. 37; Bill Nowlin, "Lou Gorman: 'You Don't Win Without Good Scouts': A GM Looks at Scouting," in *Can He Play?: A Look at Baseball Scouts and Their Profession*, ed. Jim Sandoval and Bill Nowlin, pp. 116–127 (Phoenix: Society for American Baseball Research, 2011). There are many examples of the use of the "good face"; for an explanation of the general phenomenon by longtime

scout Gary Nickels, see Kevin Kerrane, *Dollar Sign on the Muscle: The World of Baseball Scouting* (New York: Beaufort Books, 1984), pp. 234–235.

2. Art Stewart, with Sam Mellinger, *The Art of Scouting: Seven Decades Chasing Hopes and Dreams in Major League Baseball* (Olathe, KS: Ascend, 2014), p. 235.

3. "2013 SABR Analytics Conference: Bill James," YouTube video, 52:05, from a speech given at the SABR Analytics Conference in Phoenix, Arizona, on March 7, 2013, posted by SABRvideos, March 24, 2013, https: //www.youtube.com/watch?v =sbTAorEKm1c (accessed July 25, 2017).

4. Kerrane, *Dollar Sign on the Muscle*, pp. 115–116; Kerrane also prints the entire chart on p. 114.

5. Stewart with Mellinger, *Art of Scouting*, p. 245; Bill Nowlin, "Lou Gorman"; Scarborough quoted in Roger Angell, *Five Seasons: A Baseball Companion* (New York: Popular Library 1978), p. 347.

6. LaMacchia quoted in P. J. Dragseth, ed., *Eye for Talent: Interviews with Veteran Baseball Scouts* (Jefferson, NC: McFarland, 2010), p. 136; Bill Shanks, *Scout's Honor: The Bravest Way to Build a Winning Team* (New York: Sterling and Ross, 2005), pp. 120, 145.

7. Quoted in Kerrane, *Dollar Sign on the Muscle*, pp. 115–116; for the "Oriole Way": Warren Corbett, "1970 Baltimore Orioles: The Oriole Way," Society for American Baseball Research, http: //sabr.org/latest/1970-baltimore-orioles-oriole-way (accessed January 20, 2017); also excerpted in Mark Armour and Malcolm Allen, eds., *Pitching, Defense, and Three-Run Homers: The 1970 Baltimore Orioles* (Lincoln: University of Nebraska Press, 2012).

8. Byrd Douglas, *The Science of Baseball* (New York: Thomas E. Wilson & Co., 1922); Ted Williams and John Underwood, *The Science Of Hitting*, rev. ed. (New York: Simon & Schuster, 1986 [1970]); Mike Stadler, *The Psychology of Baseball: Inside the Mental Game of the Major League Player* (New York: Gotham Books, 2007).

9. Edward Marshall, "The Psychology of Baseball, Discussed by A. G. Spalding," *New York Times* (November 13, 1910), p. SM13.

10. John J. Evers and Hugh S. Fullerton, *Touching Second: The Science of Baseball* (Chicago: Reilly and Britton, 1910).

11. Ibid., pp. 250, 280–282. For Münsterberg's advocacy of experimental psychology, see Jeremy Blatter, "Screening the Psychological Laboratory: Hugo Münsterberg, Psychotechnics, and the Cinema, 1892–1916," *Science in Context* 28, no. 1 (2015): pp. 53–76; and, more broadly, idem, "The Psychotechnics of Everyday Life: Hugo Münsterberg and the Politics of Applied Psychology, 1887–1917" (Ph.D. diss., Harvard University, 2014). Even before Münsterberg, Yale's E. W. Scripture was advocating the "new psychology" as useful for athletics: C. James Goodwin, "E. W. Scripture: The Application of 'New Psychology' Methodology to Athletics," in *Psychology Gets in the Game: Sport, Mind, and Behavior, 1880–1960*, ed. Christopher D. Green and Ludy T. Benjamin, Jr., pp. 78–97 (Lincoln: University of Nebraska Press, 2009).

12. Hugh S. Fullerton, "Why Babe Ruth Is Greatest Home-Run Hitter," *Popular Science Monthly* 99, no. 4 (October 1921): pp. 19–21, 110. My analysis derives largely from Alfred H. Fuchs, "Psychology and 'The Babe,'" *Journal of the History of the Behavioral Sciences* 34, no. 2 (Spring 1998): pp. 153–165; and idem, "Psychology and

Baseball: The Testing of Babe Ruth," in Green and Benjamin, *Psychology Gets in the Game*, pp. 144–167. On Cattell, see John M. O'Donnell, *The Origins of Behaviorism: American Psychology, 1870–1920* (New York: New York University Press, 1985), pp. 31–35.

13. Fullerton, "Why Babe Ruth Is Greatest," p. 21.

14. Ibid., pp. 19, 110. On measuring aptitude, see Kurt Danziger, *Constructing the Subject: Historical Origins of Psychological Research* (Cambridge: Cambridge University Press, 1990), esp. pp. 88–100.

15. Coleman R. Griffith, "Psychology and Its Relation to Athletic Competition," *American Physical Education Review* 30, no. 4 (1925): pp. 193–199 (quotation on p. 198); O'Donnell, *Origins of Behaviorism*, p. 237. On psychology and its influential expansion into education, see Danziger, *Constructing the Subject*, esp. pp. 101–117.

16. Quotation from Griffith, "Psychology and Its Relation," p. 193. More broadly on Griffith, see Christopher D. Green, "Psychology Strikes Out: Coleman R. Griffith and the Chicago Cubs," *History of Psychology* 6, no. 3 (2003): pp. 267–283; idem, "Coleman Roberts Griffith: 'Father' of North American Sports Psychology," in Green and Benjamin, *Psychology Gets in the Game*, pp. 202–229; and Daniel Gould and Sean Pick, "Sport Psychology: The Griffith Era, 1920–1940," *Sport Psychologist* 9, no. 4 (1995): pp. 391–405 (quotation on p. 399). For Tracy, see Alan S. Kornspan, "Enhancing Performance in Sport: The Use of Hypnosis and Other Psychological Techniques in the 1950s and 1960s," in Green and Benjamin, *Psychology Gets in the Game*, pp. 253–282; and Alan S. Kornspan and Mary J. MacCracken, "The Use of Psychology in Professional Baseball: The Pioneering Work of David F. Tracy," *NINE: A Journal of Baseball History and Culture* 11, no. 2 (Spring 2003): pp. 36–43. They published their advice as Coleman R. Griffith, *Psychology and Athletics: A General Survey for Athletes and Coaches* (New York: Charles Scribner's Sons, 1928); and David F. Tracy, *The Psychologist at Bat* (New York: Sterling, 1951). Griffith's book was written a decade before his stint with the Cubs; Tracy's was written immediately after his stint with the Browns to capitalize on his relative celebrity.

17. For the lack of impact, see Green and Benjamin, *Psychology Gets in the Game*, esp. pp. 1–19. This was less the case for European sports. See John M. Hoberman, *Mortal Engines: The Science of Performance and the Dehumanization of Sport* (New York: Free Press, 1992).

18. Harold R. Raymond, C. Roy Rylander, Bruce C. Lutz, Robert C. Ziller, Paul Smith, and Robert Hannah, "The Development of Measurement Devices for the Selection and Training of Major League Personnel: Technical Report No. 1," July 27, 1962, UDARM, p. 1. The program is discussed briefly in Kerrane, *Dollar Sign on the Muscle*, pp. 127–129.

19. Raymond et al., "Development of Measurement Devices: Technical Report No. 1," pp. 2, 13–15.

20. Ibid., p. 12.

21. Harold R. Raymond, C. Roy Rylander, Bruce C. Lutz, Robert C. Ziller, Paul Smith, and Robert Hannah, "The Development of Measurement Devices for the Selection and Training of Major League Personnel: Progress Report," January 31, 1963, UDARM.

22. Ibid.

23. Kerrane, *Dollar Sign on the Muscle*, pp. 130–131; quotation from Harold R. Raymond, C. Roy Rylander, Bruce C. Lutz, Robert C. Ziller, Paul Smith, and Robert Hannah, "The Development of Measurement Devices for the Selection and Training of Major League Personnel: Progress Report No. 5," January 31, 1966, UDARM.

24. "Definitive": Jack Pastore, quoted in Kerrane, *Dollar Sign on the Muscle*, p. 132. For the Academy, see Bob Bender, "Royals' Academy Seeks Prospects," *St. Petersburg Times* (May 26, 1970), p. 2-C; Richard J. Peurzer, "The Kansas City Royals' Baseball Academy," *National Pastime* 24 (2004): pp. 3–16; Kerrane, *Dollar Sign on the Muscle*, pp. 132–136.

25. As historian Hugh Aitken summarized its importance in discussing one workers' strike against its use, the "trouble had started as soon as the stop watch made its appearance," because "in the eyes of the worker the whole Taylor system came to a focus in the personality and behavior of the man with the stop watch." Hugh G. J. Aitken, *Taylorism at Watertown Arsenal: Scientific Management in Action, 1908–1915* (Cambridge: Harvard University Press, 1960), pp. 6–7, 161. Though my examples focus on baseball, the sport was hardly alone in its appropriation of psychological tools of measurement—for an early example of psychologists' work on reaction times of Stanford football players in the 1920s, see Frank G. Baugh and Ludy T. Benjamin Jr., "An Offensive Advantage: The Football Charging Studies at Stanford University," in Green and Benjamin, *Psychology Gets in the Game*, pp. 168–201.

26. Clark quoted in Shanks, *Scout's Honor*, p. 60; "Texas Rangers Scouting Manual (Revised January 1990)," Joseph S. Branzell Collection, BA MSS 114, p. 21, BHOF; Branch Rickey, with Robert Riger, *The American Diamond: A Documentary of the Game of Baseball* (New York: Simon and Schuster, 1965), p. 78; Arthur Daley, "Automation on the Diamond," *New York Times* (March 18, 1956), pp. 214–217. For the practice of tryouts, see George Gmelch and J. J. Weiner, *In the Ballpark: The Working Lives of Baseball People* (Washington, DC: Smithsonian Institute Press, 1998), p. 30; and Kerrane, *Dollar Sign on the Muscle*, pp. 211–215.

27. Haak: Kerrane, *Dollar Sign on the Muscle*, pp. 77–78; "Texas Rangers Scouting Manual (1985)," p. 14. Tools of measurement have always required surveillance and physical discipline. In astronomy, the term "personal equation" refers to the differences each astronomer brings to measurement—one has to take into account not just the measurement, but also the human–instrument interaction that produces it. See Simon Schaffer, "Astronomers Mark Time: Discipline and the Personal Equation," *Science in Context* 2, no. 1 (1988): pp. 115–145; and Jimena Canales, *A Tenth of a Second: A History* (Chicago: University of Chicago Press, 2009), pp. 21–58.

28. Dan Gutman, *Banana Bats and Ding-Dong Balls: A Century of Unique Baseball Inventions* (New York: Macmillan, 1995), pp. 70–75; Peter Morris, *A Game of Inches: The Story Behind the Innovations that Shaped Baseball*, rev. ed. (Chicago: Dee, 2010 [2006]), pp. 371–372.

29. Cronin: Gmelch and Weiner, *In the Ballpark*, p. 31; Shannon: Daniel Okrent, *Nine Innings* (Boston: Houghton Mifflin, 1994 [1985]), p. 219; Nickels: Kerrane, *Dollar Sign on the Muscle*, p. 233; "Texas Rangers Scouting Manual (1985)," p. 14; Kachigian

quoted in Kirk Kenney, "These Three Local Baseball Scouts Have Seen It All," *San Diego Union-Tribune* (June 7, 2017), http: //www.sandiegouniontribune.com/sports /padres/sd-sp-scouts-0610-story.html (accessed June 19, 2017).

30. Kerrane, *Dollar Sign on the Muscle*, pp. 234 (Nickels) and 93 (Katalinas).

31. Dragseth, *Eye for Talent*, pp. 165–166.

32. Kerrane, *Dollar Sign on the Muscle*, p. 233.

33. Nickels: Kerrane, *Dollar Sign on the Muscle*, p. 233; technical explanation: Kevin Kerrane, *Dollar Sign on the Muscle: The World of Baseball Scouting*, rev. ed. (Prospectus Entertainment Ventures, 2013), p. v (this quotation only appears only in the 2013 edition of this work; all other references to *Dollar Sign on the Muscle* in this text refer to the 1984 edition). For an example of a report indicating the gun brand, see Cornelius, report no. 3677 (July 9, 1995), Scouting Report Collection, BHOF; "Chicago Cubs Scouting Manual," BA MSS 81, p. 23, Dan Austin Collection, Box 1, BHOF; King quoted in Kenney, "Three Local Baseball Scouts." For an example of needing to continually respond to new technologies, see Tom Tango's discussion of "Pitch Velocity: New Measurement Process, New Data Points," *Tangotiger Blog* (April 5, 2017), http: //tangotiger.com/index.php/site/article/pitch-velocity-new -measurement-process-new-data-points (accessed August 9, 2017).

34. Gib Bodet, as told to P. J. Dragseth, *Major League Scout: Twelve Thousand Baseball Games and Six Million Miles* (Jefferson, NC: McFarland, 2014), p. 117; Alexander quoted in Kerrane, *Dollar Sign on the Muscle*, pp. 41–42.

35. For Texas in the 1970s, see "1975 Scouting Manual," Joseph S. Branzell Collection, BA MSS 114, p. 21a, BHOF; for New York in the 1980s using the same form, see "New York Yankees Scouting Manual—1984," BA EPH Scouts and Scouting, BHOF; and for Kansas City in the 1990s, see "Kansas City Royals Free Agent Questionnaires," BA MSS 144, BHOF.

36. "2003–2006 Basic Agreement," roadsidephotos.sabr.org/baseball/Basic Agreement.pdf (accessed January 20, 2017); Sutton Medical History Report in "Scouting Material," Major League Scouting Bureau, 1992 Handbook, BA EPH, BHOF. The genre of sports science boosterism is well represented by Nolan Ryan and Tom House, *Nolan Ryan's Pitcher's Bible: The Ultimate Guide to Power, Precision, and Long-term Performance* (New York: Simon and Schuster, 1991).

37. Margaret A. Browne and Michael J. Mahoney, "Sport Psychology," *Annual Reviews of Psychology* 35 (1984): pp. 605–625; Gmelch and Weiner, *In the Ballpark*, p. 33. For Branzell's use of the AMI, see Boxes 2 and 4 of the Joseph S. Branzell Collection, BHOF.

38. Leland P. Lyon, "A Method for Assessing Personality Characteristics in Athletics: The Athletic Motivation Inventory" (MA thesis, California State University, San Jose, 1972), p. 11.

39. Ibid., pp. 13, 24–25, 81.

40. Ibid., pp. 69–71.

41. Thomas A. Tutko and Jack W. Richards, *Psychology of Coaching* (Boston: Allyn and Bacon, 1971); idem, *Coach's Practical Guide to Athletic Motivation: Handbook with Duplicating Masters* (Boston: Allyn and Bacon, 1972).

42. Bill Bruns, "Psychologist Tom Tutko: In Sports Winning Isn't Everything," *People* (May 13, 1974), http://www.people.com/people/archive/article/0,,20064062 ,00.html (accessed June 1, 2016). That *People* magazine would run a profile on Tutko in its first year of existence is not as surprising as it might seem, given that it was the "sister" publication of *Sports Illustrated*.

43. Stephen T. Graef, Alan S. Kornspan, and David Baker, "Paul Brown: Bringing Psychological Testing to Football," in Green and Benjamin, *Psychology Gets in the Game*, pp. 230–252; also Alan S. Kornspan, "Applying Psychology to Football in the 1930s and 1940s," *Applied Research in Coaching and Athletics Annual* 21 (2006): pp. 83–99.

44. On Dorfman and the transition in the 1980s in the use of professional psychologists, see Andrew D. Knapp and Alan S. Kornspan, "The Work of Harvey Dorfman: A Professional Baseball Mental Training Consultant," *Baseball Research Journal* 44, no. 1 (Spring 2015): pp. 27–35; and Tom Hanson and Ken Ravizza, "Issues for the Sport Psychology Professional in Baseball," in *The Psychology of Team Sports*, ed. Ronnie Lidor and Keith Page Henschen, pp. 191–215 (Morgantown, WV: Fitness Information Technology, 2003); as well as Dorfman's original book (he later wrote others more focused on specific positions), H. A. Dorfman and Karl Kuehl, *The Mental Game of Baseball: A Guide to Peak Performance* (South Bend, IN: Diamond Communications, 1989).

45. Ritterpusch quoted in Kerrane, *Dollar Sign on the Muscle*, p. 123. For the tests, see Ronald E. Smith and Donald S. Christensen, "Psychological Skills as Predictors of Performance and Survival in Professional Baseball," *Journal of Sport and Exercise Psychology* 17 (1995): pp. 399–415; Ronald E. Smith, Frank L. Smoll, Robert W. Schutz, and J. T. Ptacek, "Development and Validation of a Multidimensional Measure of Sport-Specific Psychological Skills: The Athletic Coping Skills Inventory-28," *Journal of Sport and Exercise Psychology* 17 (1995): pp. 379–398.

46. "Philadelphia Phillies Scouting Manual [1978]," BA MSS 101, p. 1-1, Tony Lucadello Collection, folder 17, BHOF.

47. Ibid., pp. 2-1, 18-3–18-6.

48. Ibid., p. 12-6; Kerrane, *Dollar Sign on the Muscle*, pp. 130–131.

49. Pages from Major League Scouting Bureau dated February 25, 1986, "Philadelphia Phillies Scouting Manual [1978]"; McLaughlin: Kerrane, *Dollar Sign on the Muscle*, p. 125; Arbuckle: Shanks, *Scout's Honor*, p. 362 (emphasis in original); Cronin: Gmelch and Weiner, *In the Ballpark*, p. 33.

50. Kerrane, *Dollar Sign on the Muscle*, pp. 123–126.

51. Cronin: Gmelch and Weiner, *In the Ballpark*, p. 33; Lennie Merullo, interview with Bob Bencks, November 10, 1994, SABR Oral History Files, https://sabr.box .com/shared/static/n9nbijp81j5hhhmn5lm1.mp3, and https://sabr.box.com/shared /static/niy5h3unnflxrw07jpwb.mp3 (accessed August 2, 2017).

52. Mock: Dragseth, *Eye for Talent*, p. 168; Gorman: Nowlin, "Lou Gorman."

53. Kerrane, *Dollar Sign on the Muscle*, p. 129. Emphasis in original.

54. Hoberman, *Mortal Engines*, pp. 187–188.

55. Griffith, "Psychology and Its Relation," p. 194; Jill G. Mirawski and Gail A. Hornstein, "Quandary of the Quacks: The Struggle for Expert Knowledge in

American Psychology, 1890–1940," in *The Estate of Social Knowledge*, ed. Joanne Brown and David Van Keuren, pp. 106–133 (Baltimore: Johns Hopkins University Press, 1991).

56. Smith and Christensen, "Psychological Skills as Predictors," pp. 412–413; and Smith et al., "Development and Validation," p. 395.

57. See, e.g., Gerd Gigerenzer, "Probabilistic Thinking and the Fight Against Subjectivity," and Kurt Danziger, "Statistical Method and the Historical Development of Research Practice in American Psychology," both in *The Probabilistic Revolution*, Vol. 2: *Ideas in the Sciences*, ed. Lorenz Krüger, Gerd Gigerenzer, and Mary S. Morgan, pp. 11–47 (Cambridge: MIT Press, 1987); Rebecca Lemov, *Database of Dreams: The Lost Quest to Catalog Humanity* (New Haven, CT: Yale University Press, 2015), esp. pp. 15–43; Peter Galison, "Image of Self," in *Things That Talk: Object Lessons from Art and Science*, ed. Lorraine Daston, pp. 257–294 (New York: Zone Books, 2004); Alicia Puglionesi, "Drawing as Instrument, Drawings as Evidence: Capturing Mental Processes with Pencil and Paper," *Medical History* 60, no. 3 (July 2016): pp. 359–387.

58. Evers and Fullerton, *Touching Second*, p. 143.

59. Nate Penn, "How to Build the Perfect Batter," *GQ* (September 2006): pp. 292–305.

60. Mann quoted in "SABR 43: Statistical Analysis Panel," YouTube video, 1:38:21, from the SABR 43 Statistical Analysis Panel in Philadelphia on Saturday, August 3, 2013, posted by SABRvideos October 25, 2013, https: //youtu.be /AO_yOYP3asw (accessed August 17, 2017); the work of Jason Sherwin and Jordan Muraskin appears in the journal *Neuroimage* (and elsewhere) and was surveyed in David Kohn, "Scientists Examine What Happens in the Brain When a Bat Tries to Meet a Ball," *Washington Post* (August 29, 2016), https: //www.washingtonpost.com /national/health-science/scientists-examine-what-happens-in-the-brain-when-bat -tries-to-meet-ball/2016/08/29/d32e9d4e-4d14-11e6-a7d8-13d06b37f256_story .html (accessed September 26, 2016); Zach Schonbrun, "As It Turns Out, Baseball Really Is Brain Science," *New York Times* (April 15, 2018), pp. SP1, SP4; for "neuroscouting": Keith Law, *Smart Baseball: The Story Behind the Old Stats that Are Ruining the Game, the New Ones that Are Running It, and the Right Way to Think about Baseball* (New York: William Morrow, 2017), pp. 263–268; "wearables": Kate Crawford, Jessa Lingel, and Tero Karppi, "Our Metrics, Ourselves: A Hundred Years of Self-tracking from the Weight Scale to the Wrist Wearable Device," *European Journal of Cultural Studies* 18, nos. 4–5 (2015): pp. 479–496.

61. Penn, "How to Build the Perfect Batter," p. 297; Hannah quoted in Kerrane, *Dollar Sign on the Muscle*, p. 129.

62. Davis quoted in Kerrane, *Dollar Sign on the Muscle*, p. 143 (emphasis in original); Hoberman, *Mortal Engines*, esp. p. 32, notes that while the idea of the measured body extends back more than a century (Adolphe Quetelet's 1835 *Sur L'homme et le Développement de ses Facultés* is usually credited with popularizing the concept of the mathematically "average man"), the practice of bodily measurement to determine "types" became widespread in the mid-twentieth century; e.g., Anna G. Creadick, *Perfectly Average: The Pursuit of Normality in Postwar America*

(Amherst: University of Massachusetts Press, 2010). Of course, scouts were pursuing outliers—extraordinary physical attributes—rather than normality, but the same tools were used.

Chapter 7: A Machine for Objectivity

1. Rossi: Biggio, report no. 5360 (March 16, 1987); Kohler: Biggio, report no. 1236 (March 15, 1987); Labossiere: Biggio, report no. 1033 (July 1, 1986), all from Scouting Report Collection, BHOF.

2. "Washington Senators Scouts Manual (1961)," Joseph S. Branzell Collection, Box 4, p. 14, BHOF.

3. Ibid., pp. 14–15.

4. Ibid., p. 34. For Branzell, see P. J. Dragseth, *Major League Baseball Scouts: A Biographical Dictionary* (Jefferson, NC: McFarland, 2011), p. 46.

5. "Washington Senators Scouts Manual (1961)," pp. 7, 21.

6. Ibid., p. 9.

7. Branzell appears to have amended the original 1–7 scale with a 1–5 scale (1 being excellent, 5 being poor); the example card found in ibid., p. 21, has no grade other than 1–5, so it is unclear if Washington's scouts used different systems simultaneously in 1961.

8. On the draft, see W. C. Madden, *Baseball's First-Year Player Draft, Team by Team Through 1999* (Jefferson, NC: McFarland, 2001); and Allan Simpson, ed., *The Baseball Draft: The First 25 Years, 1965–1989* (Durham, NC: American Sports Publishing, 1990).

9. Simpson, *Baseball Draft*, p. 18.

10. Quoted in ibid., p. 18 Though more clubs initially opposed the draft, only one ultimately voted against it.

11. Ibid., p. 19.

12. P. J. Dragseth, ed., *Eye for Talent: Interviews with Veteran Baseball Scouts* (Jefferson, NC: McFarland, 2010), pp. 64 (Clary), 46 (Gardner), 111 (Thompson), 206 (Rizzo).

13. Brown quoted in Simpson, *Baseball Draft*, p. 17; Alexander quoted in Dragseth, *Eye for Talent*, p. 23.

14. "Philadelphia Phillies Scouting Manual [1978]," BA MSS 101, Tony Lucadello Collection, folder 17, p. 3-1, BHOF.

15. Jack Lang, "Fanning Tries to Breathe Life into Central Scouting Bureau," *Sporting News* (February 17, 1968), p. 23; C. C. Johnson Spink, "Gallagher Proposed Scouting Plan," *Sporting News* (February 13, 1965), p. 12; Branch Rickey with Robert Riger, *The American Diamond: A Documentary of the Game of Baseball* (New York: Simon and Schuster, 1965), pp. 200–202.

16. Lang, "Fanning Tries to Breathe Life"; "Scouting Bureau Seeks a Successor to Fanning," *Sporting News* (August 31, 1968), p. 7; Norm King, "Jim Fanning," Society for American Baseball Research, http://sabr.org/bioproj/person/feaf120c (accessed August 17, 2017).

17. For a postmortem, see Fanning's comments in Stan Isle, "High Costs Revive Central Scouting Concept," *Sporting News* (July 20, 1974), p. 43.

18. "Caught on the Fly," *Sporting News* (March 9, 1974), p. 47; for the bureau more generally, see also "Wilson to Head Central Scouting Bureau," *Sporting News* (August 17, 1974), p. 22; and Lowell Reidenbaugh, "Brown Hails Central Scouting as Plus for All," *Sporting News* (October 12, 1974), pp. 17–18.

19. Lou Chapman, "New Bureau Chief Wilson Maps Central Scouting Plan," *Sporting News* (August 24, 1974), p. 23; "Texas Rangers Scout's Manual (1975)," Branzell Collection, box 4, p. 1, BHOF. This account was confirmed by an early bureau scout, Lennie Merullo, in his interview with Bob Bencks, November 10, 1994, in SABR Oral History Files, https: //sabr.box.com/shared/static/n9nbijp81j5hhhmn5lml.mp3, and https: //sabr.box.com/shared/static/niy5h3unnflxrw07jpwb.mp3 (accessed August 2, 2017).

20. "Central Scouting Appeal," *Sporting News* (December 4, 1976), p. 71; "Caught on the Fly," p. 47; Murray Chass, "Centralized Scouting System Loses Grip," *Sporting News* (April 8, 1978), p. 30; "National League Flashes," *Sporting News* (October 18, 1980), p. 46; Ken Nigro, "Scouting Bureau Chugs Along on Rocky Road," *Sporting News* (September 3, 1977), p. 11.

21. Dave Nightingale, "Who'll Follow Kuhn?: The Search Goes On," *Sporting News* (December 19, 1983), p. 47; Kevin Kerrane, *Dollar Sign on the Muscle: The World of Baseball Scouting* (New York: Beaufort Books, 1984), p. 297.

22. Milwaukee: Daniel Okrent, *Nine Innings* (Boston: Houghton Mifflin, 1994 [1985]), pp. 220–224; St. Louis: Dragseth, *Eye for Talent*, p. 184 (OFP of 40+ was specified); for resentment: Dragseth, *Eye for Talent*, p. 120; Bill Fleischman, "Majors Cut 168 in Scouting Staffs," *Sporting News* (March 15, 1975), p. 55; Kerrane gives an estimate of 250 scouts fired in *Dollar Sign on the Muscle*, p. 21, though he offers similar financial estimates on pp. 22–24.

23. "Texas Rangers Scout's Manual (1975)," pp. D-1 [title page] and 2.

24. "Philadelphia Phillies Scouting Manual [1978]," p. 3-1; "New York Yankees Scouting Manual," BA EPH, Box 1, pp. 1–2, BHOF.

25. "Pittsburgh Pirates [1949] Operations Manual," MFF 525, p. 5, BHOF.

26. Kerrane, *Dollar Sign on the Muscle*, p. 18.

27. "Philadelphia Phillies Scouting Manual [1978]," pp. 4-1 and 4-2 (noting the change was made in 1972); on Davis's role, see Kerrane, *Dollar Sign on the Muscle*, p. 148.

28. Kerrane, *Dollar Sign on the Muscle*, p. 18. Emphasis in original.

29. "Texas Rangers Scout's Manual (1975)," p. 7.

30. Cronin: George Gmelch and J. J. Weiner, *In the Ballpark: The Working Lives of Baseball People* (Washington, DC: Smithsonian Institute Press, 1998), p. 32; Gorman: Bill Nowlin, "Lou Gorman: 'You Don't Win Without Good Scouts': A GM Looks at Scouting," in *Can He Play? A Look at Baseball Scouts and their Profession*, ed. Jim Sandoval and Bill Nowlin, pp. 116–127 (Phoenix: Society for American Baseball Research, 2011); Haak: Kerrane, *Dollar Sign on the Muscle*, p. 73.

31. "Philadelphia Phillies Scouting Manual [1978]," p. 4-2.

32. This January 27, 1975, letter located in "Texas Rangers Scout's Manual (1975)," after p. 7.

33. "Texas Rangers Scout's Manual (1975)," pp. 10A–10F.

34. Pages from the Major League Scouting Bureau dated February 25, 1986, and located in folder 17, BA MSS 101, Tony Lucadello Collection, BHOF.

35. "Texas Rangers Scout's Manual (1975)," pp. 10A–10F; "New York Yankees Scouting Manual," pp. 7–8.

36. "Texas Rangers Scout's Manual (1975)," pp. 10A–10F.

37. Interview with Don Pries, November 20, 2017.

38. "Texas Rangers Scout's Manual (1975)," p. 10; "Rangers Scouting Manual (Revised January, 1985)," BA MSS 114, Joseph S. Branzell Collection, Box 4, pp. 6, 24–25, BHOF.

39. See, e.g., Mauriello, report no. 1546 (March 1952); Bannister, report no. 1657 (March 17, 1969), both in Scouting Report Collection, BHOF.

40. See, e.g., Dauer, report no. 1688 (March 27, 1974); Burtt, report no. 1617 (March 1, 1975), both in Scouting Report Collection, BHOF.

41. Blomberg, report no. 5051 (April 12, 1967); Hurdle, report no. 5156 (March 31, 1975), both in Scouting Report Collection, BHOF.

42. Minnesota: e.g., Hrbek, report no. 1466 (May 2, 1978); Cincinnati: e.g., Esasky, report no. 1073 (May 2, 1978); Cubs: Biggio, report no. 31 (March 17, 1987), all in Scouting Report Collection, BHOF.

43. Collins: Kerrane, *Dollar Sign on the Muscle*, p. 253; Gib Bodet, as told to P. J. Dragseth, *Major League Scout: Twelve Thousand Baseball Games and Six Million Miles* (Jefferson, NC: McFarland, 2014), p. 138.

44. Mock: Dragseth, *Eye for Talent*, p. 168; Bodet, *Major League Scout*, pp. 27, 52.

45. "Philadelphia Phillies Scouting Manual [1978]," p. 12-3.

46. "Grading System," in "Chicago Cubs Scouting Manual," BA MSS 81, Dan Austin Collection, Box 1, BHOF.

47. Ibid.

48. Frank McGowan, "Judgment—The First Ingredient in Scouting," BA MSS 114, Box 1, folder 9, p. 1, Joseph S. Branzell Collection, BHOF.

49. "Texas Rangers Scout's Manual (1975)"; "Philadelphia Phillies Scouting Manual [1978]," p. 12-3; "Texas Rangers Scouting Manual (Revised January 1990)," BA MSS 114, p. 13, Joseph S. Branzell Collection, BHOF (emphasis in original).

50. "New York Yankees Scouting Manual"; "Philadelphia Phillies Scouting Manual [1978]," p. 3-1; "Why You Are Here," in "Major League Scouting Bureau Scout Development Program, Arizona, 2012," BA MSS 180, p. 4, BHOF.

51. For one view of a cross-checker on this process, see Dragseth, *Eye for Talent*, p. 119; John Stokoe's experience cross-checking for Baltimore was similar—see Paul Post, *Foresight 20/20: The Life of Baseball Scout Ralph DiLullo* (self-published, 1995), pp. 58–61 (a copy is deposited at BHOF). It is unclear who pioneered the use of cross-checkers. Kerrane credits McLaughlin in 1955, while Don Pries credits Walter Shannon in 1960: Kerrane, *Dollar Sign on the Muscle*, p. 117; and Don Pries, *A Father's Baseball Dream Becomes a Son's Journey* (Bloomington, IN: WestBow Press, 2016), ch. 17.

52. Gib Bodet, "National Cross Checking," in Sandoval and Nowlin, *Can He Play?*, pp. 150–156.

53. Bodet, *Major League Scout*, pp. 50–51; Art Stewart, with Sam Mellinger, *The Art of Scouting: Seven Decades Chasing Hopes and Dreams in Major League Baseball* (Olathe, KS: Ascend, 2014), p. 239.

54. Cronin: Gmelch and Weiner, *In the Ballpark*, p. 35.

55. King quoted in Kirk Kenney, "These Three Local Baseball Scouts Have Seen It All," *San Diego Union-Tribune* (June 7, 2017), http://www.sandiegouniontribune.com/sports/padres/sd-sp-scouts-0610-story.html (accessed June 19, 2017); Poitevint: Okrent, *Nine Innings*, p. 219; Melvin: George Sullivan, *Baseball Backstage* (New York: Holt, Rinehart, and Winston, 1986), pp. 15–16; Bodet, *Major League Scout*, pp. 16, 50; "Philadelphia Phillies Scouting Manual [1978]," pp. 1-1, 12-7, and 18-2.

56. Bodet: Dragseth, *Eye for Talent*, pp. 191–192; and Bodet, *Major League Scout*, p. 85; "tipping hand": Gmelch and Weiner, *In the Ballpark*, pp. 28–34; Stewart, *Art of Scouting*, p. 233.

57. Robert Sawyer, "Gene Bennett: 'My Goal Is to Give Back to the Game as Much as It Has Given to Me," in Sandoval and Nowlin, *Can He Play?*, pp. 134–140; see also "Reds Announce Retirement of Gene Bennett," press release (January 27, 2011), http://cincinnati.reds.mlb.com/news/press_releases/press_release.jsp?ymd=20110127&content_id=16520254&fext=.jsp&c_id=cin&vkey=pr_cin (accessed January 23, 2017); "Philadelphia Phillies Scouting Manual [1978]," p. 12-1.

58. Dragseth, *Eye for Talent*, pp. 162 (Mock), 31–32 (Major League Scouting Bureau), 176 (McAlister).

59. Stewart, *Art of Scouting*, pp. 164–167.

60. Kerrane, *Dollar Sign on the Muscle*, p. 119.

61. Nowlin, "Lou Gorman."

62. Ibid.

63. Clary: Dragseth, *Eye for Talent*, p. 68; "Scouting Material," *Major League Scouting Bureau Handbook* (1992), BA EPH, p. 6, BHOF; Schwartz: "Lefebvre to Manage," *Sporting News* (March 11, 1978), p. 63.

64. "Masters of the noncommittal" was Okrent's description, based on talking with Poitevint, in Okrent, *Nine Innings*, p. 220; Brown: Joe McGuff, "Royals Cut Scout Staff, Farm System," *Sporting News* (November 23, 1974), p. 61; Nieman: Russell Schneider, "Nieman Blasts Central Scouting Proposal," *Sporting News* (August 3, 1974), p. 20; Merullo, interview with Bencks, November 10, 1994.

65. Haak quoted in Kerrane, *Dollar Sign on the Muscle*, p. 84; descriptions of bureau procedures are in "Scouting Material," pp. 21–22. Other midcentury industries similarly worked to find ways of rendering subjective evaluations—regarding, e.g., the taste of wine or the experience of pain—as useful, quantified judgments. See, e.g., the paired articles Steven Shapin, "A Taste of Science: Making the Subjective Objective in the California Wine World," *Social Studies of Science* 46, no. 3 (2016): pp. 436–460; and Christopher J. Phillips, "The Taste Machine: Sense, Subjectivity, and Statistics in the California Wine-World," *Social Studies of Science* 46, no. 3 (2016): pp. 461–481. Feminist scholars were some of the first to emphasize that the idea of a

"view from nowhere" was always misleading; see, e.g., Donna Harraway, "Situated Knowledges: The Science Question in Feminism and the Privilege of Partial Perspective," *Feminist Studies* 14, no. 3 (1988): pp. 575–599.

66. For one perspective on the OFP's continued use, see Ben Jedlovec, "An Intern's Perspective on the Draft," in Sandoval and Nowlin, *Can He Play?*, pp. 113–116. There is a close historical connection between the collection of data, particularly numerical data, and the "reduction" of people to—or "replacement" of people with—such data. For the classic sense in which these kinds of data-keeping create categories of people, see Ian Hacking, "Making Up People," in *The Science Studies Reader*, ed. Mario Biagioli, pp. 161–171 (New York: Routledge, 1999 [1986]). In the twentieth century, insurance ratings, public opinion polling, credit scores, and risk factors all constitute powerful and useful representations of people with numbers. Respectively, see Dan Bouk, *How Our Days Became Numbered: Risk and the Rise of the Statistical Individual* (Chicago: University of Chicago Press, 2015); Sarah E. Igo, *The Averaged American: Surveys, Citizens, and the Making of a Mass Public* (Cambridge: Harvard University Press, 2007); Louis Hyman, *Debtor Nation: The History of America in Red Ink* (Princeton, NJ: Princeton University Press, 2011); Jeremy A. Greene, *Prescribing by Numbers: Drugs and the Definition of Disease* (Baltimore: Johns Hopkins University Press, 2007); and William G. Rothstein, *Public Health and the Risk Factor: A History of an Uneven Medical Revolution* (Rochester, NY: University of Rochester Press, 2003).

67. For a different perspective, see Theodore M. Porter, *Trust in Numbers: The Pursuit of Objectivity in Science and Public Life* (Princeton, NJ: Princeton University Press, 1995), esp. pp. 200–216; and idem, "Objectivity as Standardization: The Rhetoric of Impersonality in Measurement, Statistics, and Cost-Benefit Analysis," *Annals of Scholarship* 9, nos. 1–2 (1992): pp. 19–60.

68. On the bureau's efforts from the 1970s to the 1990s, see Lowell Reidenbaugh, "Brown Hails Central Scouting as Plus for All," *Sporting News* (October 12, 1974), pp. 17–18; and "Scouting Material," p. 20. For more recent restructuring, see Michael Lananna, "Major League Scouting Bureau Restructuring Under Bavasi," *Baseball America* (January 14, 2016), http: //www.baseballamerica.com/draft/major-league-scouting-bureau-restructuring-bavasi/#ucoY0Hkl8LsHcHr5.97 (accessed January 23, 2017); Chad Jennings, "Nightengale: For MLB Scouts, It's Getting Harder to Avoid Force-out," *Lohud: The Journal News* (January 16, 2016), http: //www.lohud.com/story/sports /mlb/lohud-yankees/2016/01/16/nightengale-mlb-scouts-getting-harder-avoid-force /78901856/ (accessed January 23, 2017); and for another take, Rob Neyer, "Baseball's Scouting Profession Isn't Downsizing, It's Just Evolving," *Fox Sports* (January 28, 2016), http: //www.foxsports.com/mlb/story/mlb-scouts-trend-sabermetrics-analysis -experience-franchise-downsize-012816 (accessed January 23, 2017).

69. "Scouting Material," pp. 4–14 and unnumbered pages: "Position Player Phrases" and "Pitcher Phrases"; see also Bernie Pleskoff, "School Equips Aspiring Scouts," *mlb.com* (July 6, 2011), http: //mlb.mlb.com/news/print.jsp?ymd=20110706 &content_id=21492424&c_id=mlb (accessed January 23, 2017); Ben Lindberg, "Welcome to Baseball Scout School," *Grantland* (October 7, 2013), http: //grantland.com /the-triangle/welcome-to-baseball-scout-school/ (accessed January 23, 2017).

70. "Philadelphia Phillies Scouting Manual [1978]," p. 1-3; "Grading System," pp. 1–15, in "Chicago Cubs Scouting Manual," BHOF.

71. Stewart, *Art of Scouting*, p. 239; Bodet, *Major League Scout*, p. 23.

72. Simpson, *Baseball Draft*, p. 19. Gillick likely meant "exact science," but the point is the same.

73. These descriptions of "data-driven" sciences are taken from David Sepkoski, "Towards 'A Natural History of Data': Evolving Practices and Epistemologies of Data in Paleontology, 1800–2000," *Journal of the History of Biology* 46, no. 3 (2013): pp. 401–444; and S. Leonelli, "Introduction: Making Sense of Data-Driven Research in the Biological and Biomedical Sciences," *Studies in History and Philosophy of Biological and Biomedical Sciences* 43 (2012): pp. 1–3. There are, of course, many differences in data-driven sciences after the advent of electronic computing compared to before, even as parallels with traditional data-driven practices like botany remain: Bruno J. Strasser, "Data-Driven Sciences: From Wonder Cabinets to Electronic Databases," *Studies in History and Philosophy of Biological and Biomedical Sciences* 43 (2012): pp. 85–87.

74. Katalinas: Kerrane, *Dollar Sign on the Muscle*, 239; Dragseth, *Eye for Talent*, pp. 120 (Mattick) and 179 (McAlister).

Conclusion

1. Rebecca Lemov, "Big Data Is People!" *Aeon* (June 16, 2016), https://aeon.co/essays/why-big-data-is-actually-small-personal-and-very-human (accessed January 24, 2017).

2. Erik Malinowski, *Betaball: How Silicon Valley and Science Built One of the Greatest Basketball Teams in History* (New York: Atria, 2017); Duncan Alexander, *Outside the Box: A Statistical Journey Through the History of Football* (London: Random House, 2017); Wayne L. Winston, *Mathletics: How Gamblers, Managers, and Sports Enthusiasts Use Mathematics in Baseball, Basketball, and Football* (Princeton, NJ: Princeton University Press, 2012); Mike Leach and Ferhat Guven, eds., *Sports for Dorks: College Football* (Houston: Sports Dorks CFB, 2011).

3. Allan Simpson, ed., *The Baseball Draft: The First 25 Years, 1965–1989* (Durham, NC: American Sports Publishing, 1990), p. 19; for more recent data, see Richard T. Karchar, "The Chances of a Drafted Baseball Player Making the Major Leagues: A Quantitative Study," *Baseball Research Journal* 46, no. 1 (2017): pp. 50–55, https://sabr.org/research/chances-drafted-baseball-player-making-major-leagues-quantitative-study (accessed April 18, 2018).

4. For a range of critical approaches, see Bruno J. Strasser and Paul N. Edwards, "Big Data Is the Answer ... But What Is the Question?" *Osiris* 32 (2017): pp. 328–354; Brian Beaton, Amelia Acker, Lauren Di Monte, Shivrang Setlur, Tonia Sutherland, and Sarah E. Tracy, "Debating Data Science: A Roundtable," *Radical History Review* 127 (January 2017): pp. 133–148; Rob Kitchin, *The Data Revolution: Big Data, Open Data, Data Infrastructures and Their Consequences* (London: Sage, 2014); danah boyd and Kate Crawford, "Critical Questions for Big Data: Provocations for a Cultural,

Technological, and Scholarly Phenomenon," *Information, Communication & Society* 15, no. 5 (2012): pp. 662–679.

5. Cf. Theodore M. Porter, "Objectivity as Standardization: The Rhetoric of Impersonality in Measurement, Statistics, and Cost-Benefit Analysis," *Annals of Scholarship* 9, nos. 1–2 (1992): pp. 19–60. One reason scouting and scoring might have initially retained an element of judgment despite pressures to the contrary was that for most of the twentieth century the practices were limited to a relatively small community, analogous to the ways in which measurements of space and quantity might have once been greatly variable, only to be standardized as economies expanded across regions and borders. For a classic example of the work of standardizing measurements, see Witold Kula, *Measures and Men*, trans. R. Szreter (Princeton, NJ: Princeton University Press, 1986).

6. These numbers are taken from baseball-almanac.com. It has long been the case that both facts and quantification have roots in both the needs of states to account for subjects and mercantile exchanges, with questions of fairness, commensurability, and equivalence being central to thinking about the rise of numbers and different forms of calculation. For a sample of this vast literature, see Gerg Gigerenzer, Zeno Swijtink, Theodore Porter, Lorraine Daston, John Beatty, and Lorenz Kruger, *The Empire of Chance: How Probability Changed Science and Everyday Life* (Cambridge: Cambridge University Press, 1989); Mary Poovey, *A History of the Modern Fact* (Chicago: University of Chicago Press, 1998); Lorraine Daston, *Classical Probability and the Enlightenment* (Princeton, NJ: Princeton University Press, 1988); Theodore M. Porter, *Trust in Numbers: The Pursuit of Objectivity in Science and Public Life* (Princeton, NJ: Princeton University Press, 1995); and, for a different perspective, Alain Desrosières, *The Politics of Large Numbers: A History of Statistical Reasoning* (Cambridge: Harvard University Press, 1998 [1993]).

7. For a historical perspective, see Ann Blair, *Too Much to Know: Managing Scholarly Information Before the Modern Age* (New Haven, CT: Yale University Press, 2010); and for overload within scientific disciplines specifically, see Staffan Müller-Wille and Isabelle Charmantier, "Natural History and Information Overload: The Case of Linnaeus," *Studies in History and Philosophy of Biological and Biomedical Sciences* 43 (2012): pp. 4–15; Peter Keating and Alberto Cambrosio, "Too Many Numbers: Microarrays in Clinical Cancer Research," *Studies in History and Philosophy of Biological and Biomedical Sciences* 43 (2012): pp. 37–51; and Brian W. Ogilvie, "The Many Books of Nature: Renaissance Naturalists and Information Overload," *Journal of the History of Ideas* 64, no. 1 (January 2003): pp. 29–40. Neither scouts nor scorers complained of "overload," though many did note quantification was increasing and that ever more importance was given to those numbers.

8. Jon Agar, "What Difference Did Computers Make?" *Social Studies of Science* 36, no. 6 (2006): pp. 869–907; see also idem, *The Government Machine: A Revolutionary History of the Computer* (Cambridge: MIT Press, 2003); and Elena Aronova, Christine von Oertzen, and David Sepkoski, "Introduction: Historicizing Big Data," *Osiris* 32 (2017): pp. 1–17 (quotation on p. 16). Cf. Hallam Stevens, "A Feeling for the Algorithm: Working Knowledge and Big Data in Biology," *Osiris* 32 (2017):

pp. 151–174; idem, *Life Out of Sequence: A Data-Driven History of Bioinformatics* (Chicago: University of Chicago Press, 2013); and Stephanie Aleen Dick, "After Math: (Re)configuring Minds, Proof, and Computing in the Postwar United States" (Ph.D. diss., Harvard University, 2015).

9. For two recent studies, see Steven Connor, *Living by Numbers: In Defense of Quantity* (London: Reaktion Books, 2016); and Robert Tubbs, *Mathematics in Twentieth-Century Literature and Art: Content, Form, Meaning* (Baltimore: Johns Hopkins University Press, 2014).

10. Tom Nichols, *The Death of Expertise: The Campaign Against Established Knowledge and Why It Matters* (Oxford: Oxford University Press, 2017); Viktor Mayer-Schönberger and Kenneth Cukier, *Big Data: A Revolution that Will Transform How We Live, Work, and Think* (Boston: Houghton Mifflin, 2013), pp. 139–145. For Pearson, see J. Rosser Matthews, "Almroth Wright, Vaccine Therapy and British Biometrics: Disciplinary Expertise Versus Statistical Objectivity," in *The Road to Medical Statistics*, ed. Eileen Magnello and Anne Hardy, pp. 125–147 (Amsterdam: Rodopi, 2002). An introduction to different conceptions of expertise and how they are acquired is Harry Collins and Robert Evans, *Rethinking Expertise* (Chicago: University of Chicago Press, 2007). For academic debates surrounding the issue, see Harry Collins, Robert Evans, and Martin Weinel, "STS as Science or Politics?" *Social Studies of Science* 47, no. 4 (2017): pp. 580–586; Michael Lynch, "STS, Symmetry, and Post-Truth," *Social Studies of Science* 47, no. 4 (2017): pp. 593–599; and Sheila Jasanoff and Hilton R. Simmet, "No Funeral Bells: Public Reason in a 'Post-Truth' Age," *Social Studies of Science* 47, no. 5 (2017): pp. 751–770 in response to Sergio Sismondo, "Post-Truth?" *Social Studies of Science* 47, no. 1 (2017): pp. 3–6.

11. For voting fraud controversies, see Stephen Ansolabehere, Samantha Luks and Brian Schaffner, "Trump Wants to Investigate Purported Mass Voter Fraud: We Pre-debunked His Evidence," *Washington Post* (October 19, 2016), https://www.washingtonpost.com/news/monkey-cage/wp/2016/10/19/trump-thinks-non-citizens-are-deciding-elections-we-debunked-the-research-hes-citing/ (accessed July 31, 2017); and for the original article in question, see Jesse T. Richman, Gulshan A. Chattha, and David C. Earnest, "Do Non-citizens Vote in U.S. Elections?" *Electoral Studies* 36 (December 2014): pp. 149–157. For challenging of expert consensus: Naomi Oreskes and Erik M. Conway, *Merchants of Doubt: How a Handful of Scientists Obscured the Truth on Issues from Tobacco Smoke to Global Warming* (New York: Bloomsbury, 2010). Among many studies of p-hacking, see Megan L. Head, Luke Holman, Rob Lanfear, Andrew T. Kahn, and Michael D. Jennions, "The Extent and Consequences of P-Hacking in Science," *PLoS Biology* 13, no. 3 (2015): e1002106, https://doi.org/10.1371/journal.pbio.1002106 (accessed July 31, 2017); and John P. A. Ioannidis, "Why Most Published Research Findings Are False," *PLoS Medicine* 2(8): e124, https://doi.org/10.1371/journal.pmed.0020124 (accessed July 31, 2017). For an alternative (and imaginative) view of statistics as *hindering* the spread of knowledge, see Naomi Oreskes and Erik M. Conway, *The Collapse of Western Civilization: A View from the Future* (New York: Columbia University Press, 2014), p. 17; an introduction to the large, and growing, literature on statistics and causality

is Judea Pearl, "Causal Inference in Statistics: An Overview," *Statistics Surveys* 3 (2009): pp. 96–146.

12. Bill James, *The Bill James Baseball Abstract 1982* (New York: Ballantine Books, 1982), p. 28.

13. Mayer-Schönberger and Cukier, *Big Data*, p. 143; Lewis makes the distinction most clearly in the afterword to the 2004 edition of his book: Michael Lewis, *Moneyball: The Art of Winning an Unfair Game* (New York: Norton, 2004 [2003]), p. 287; Jim Rutenberg, "'Alternative Facts' and the Costs of Trump-Based Reality," *New York Times* (January 23, 2017), https: //www.nytimes.com/2017/01/22/business/media /alternative-facts-trump-brand.html (accessed August 1, 2017); Oxford Dictionaries, "2016 International Word of the Year," press release, https: //www.oxforddictionaries .com/press/news/2016/12/11/WOTY-16 (accessed September 27, 2017).

INDEX

objectivity: of data, 1, 3; and expertise, 248–49; history of, 7, 260n8; in scoring, 6, 57–58, 64, 84–86, 94–96, 245; in scouting, 178–81, 198–99, 201, 227, 231, 233–37, 240, 242

observation. *See* vision

Official Encyclopedia of Baseball (Turkin and Thompson), 19, 21–22

official scorer: and appeal process, 88–91, 93–94; appointment of, 62–63, 70, 76–77; controversy concerning, 62–64, 72–77, 88–91; credibility of, 37; location on field of, 7–8, 37, 87; and relationship with clubs, 63, 95; report of, 20, 31, 60–61, 67, 250; in rulebook, 77–78; vs. umpires, 59–62, 94; vs. unofficial scoring, 69, 75–76. *See also* scoring

Ogilvie, Bruce, 187–88

overall future potential (OFP), 3, 200–201, 219–27, 230, 236–38, 253

Palmer, Pete: database of, 18–32, 108–9, 117, 249; employment of, 19, 24–25, 27; and Total Sports, 125; on Ty Cobb's record, 73

Pastore, Jack, 192

Paul, Gabe, 306

Pearson, Karl, 247, 259n1

Plaschke, Bill, 62

Poitevint, Ray, 231

Pries, Don, 210–12, 220, 238

Project Scoresheet, 99–100, 103–8, 117, 124–26, 247; legacy of, 108, 111, 118–19

psychology. *See* science

Pujols, Albert, 197

punch cards (Hollerith cards), 22–23, 31, 119, 180

Quetelet, Adolphe, 51, 259n1, 267n35, 287n62

race: of scorers 7; of scouts 7, 154–55, 160

radar gun, 183–86. *See also* technology

Raymond, Harold, 178–81

Raytheon, 19, 27

replay. *See* official scorer; vision

reserve clause, 143, 165–66. *See also* free agency

Retrosheet, 111–18, 249

Rickey, Branch, 144, 182, 208, 215

Ritterpusch, Dave, 191, 193, 198

Rizzo, Phil, 153, 207

Rossi, Phil, 200

Ruane, Tom, 115–17

Ruth, Babe, 175–77, 197

sabermetrics, 12–13, 102, 170–72, 242, 248

salary: of official scorers, 63, 74, 78, 80; of players 245–46; of scouts, 154

Sanborn, I. E., 84–86

Schwartz, Cory, 127–35, 249

Schwartz, Jack, 235–36

science: vs. art of scoring, 93; and baseball, 8, 51; data-based vs. human-based, 243; and data collection, 30–31, 264n26; exact vs. inexact, 31; of nutrition and scouting, 186–87; of psychology and scouting, 170–71, 174–82, 187–91, 195–99; scouting as, 140, 169, 174, 186, 191–92, 235, 241–43; as useful for scoring, 99, 108, 111–12, 116–17, 119, 134–35, 243; as useful for scouting, 195–96

Scioscia, Mike, 222–23

scorebook: commercial sale of, 44–45, 55–56; as historical record, 112–13; origins of, 7, 38–40

scoresheet: design of, 102–3, 121; early use of, 66; as technology, 8; uniformity of, 44–45

scoring: as affects gameplay, 48–50; for awarding of prizes, 43–44; difficulty of, 65–66; as double-entry accounting, 50–51; as internal recordkeeping, 40–41; origins of, 7–8, 36–41; practice of, 45–46, 55–56, 67–68, 85–86, 121–22; rules for, 37, 48–49; variations in, 55–56. *See also* official scorer

scout school. *See* training

scouting: business of, 146–47, 202–5; and control of labor, 141–42, 165–66; etymology of, 140, 161; jargon of, 155–57; and physical appearance, 203–4; practice of as compared to

A NOTE ON THE TYPE

This book has been composed in Adobe Text and Gotham.
Adobe Text, designed by Robert Slimbach for Adobe,
bridges the gap between fifteenth- and sixteenth-century
calligraphic and eighteenth-century Modern styles.
Gotham, inspired by New York street signs, was designed
by Tobias Frere-Jones for Hoefler & Co.